Advances in Science, Technology & Innovation (ASTI) is a series of peer-reviewed books based on the best studies on emerging research that redefines existing disciplinary boundaries in science, technology and innovation (STI) in order to develop integrated concepts for sustainable development. The series is mainly based on the best research papers from various IEREK and other international conferences, and is intended to promote the creation and development of viable solutions for a sustainable future and a positive societal transformation with the help of integrated and innovative science-based approaches. Offering interdisciplinary coverage, the series presents innovative approaches and highlights how they can best support both the economic and sustainable development for the welfare of all societies. In particular, the series includes conceptual and empirical contributions from different interrelated fields of science, technology and innovation that focus on providing practical solutions to ensure food, water and energy security. It also presents new case studies offering concrete examples of how to resolve sustainable urbanization and environmental issues. The series is addressed to professionals in research and teaching, consultancies and industry, and government and international organizations. Published in collaboration with IEREK, the ASTI series will acquaint readers with essential new studies in STI for sustainable development.

More information about this series at http://www.springer.com/series/15883

Advances in Science, Technology & Innovation

IEREK Interdisciplinary Series for Sustainable Development

Rashid A. Khan • Hassan Qudrat-Ullah

Adoption of LMS in Higher Educational Institutions of the Middle East

Rashid A. Khan
DCC—King Fahd University of Petroleum
and Minerals
Dhahran, Saudi Arabia

Hassan Qudrat-Ullah
KBS—King Fahd University of Petroleum
and Minerals
Dhahran, Saudi Arabia

ISSN 2522-8714   ISSN 2522-8722 (electronic)
Advances in Science, Technology & Innovation
IEREK Interdisciplinary Series for Sustainable Development
ISBN 978-3-030-50114-3   ISBN 978-3-030-50112-9 (eBook)
https://doi.org/10.1007/978-3-030-50112-9

This Springer imprint is published by the registered company Springer Nature Switzerland AG
The registered company address is: Gewerbestrasse 11, 6330 Cham, Switzerland

Dedication

First and foremost, I would like to thank God, for the strength, power, and all the blessings to complete this work. I would like to thank my family who has sacrificed their precious time and efforts to support me in the work; their encouragement and never-ending flow of support had given me the confidence I needed.

Rashid A. Khan

To my teachers who played a critical role in shaping and building my life:

–Sufi Fazal Hussain-(late)—Government Primary (now Middle) school, Mandi Bhalwal, Gujrat, Pakistan
–M. Yonus—Government Middle (now Higher Secondary) school, Thil, Gujrat, Pakistan
–M. Khalil—Government Middle (now Higher Secondary) school, Thil, Gujrat, Pakistan
–Abdul Malik—Government Middle (now Higher Secondary) school, Thil, Gujrat, Pakistan
–M. Mushtaq—Government Middle (now Higher Secondary) school, Thil, Gujrat, Pakistan
–M. Kamal-(late)—Government M.C. Islamia High (now Higher Secondary) school, Jhelum, Pakistan
–C. M. Sadiq-(late)—Government M.C. Islamia High (now Higher Secondary) school, Jhelum, Pakistan
–Prof. Azmat Chaudhry—Government College, G.T. Road & Government Degree College, Tahinianwala, Jhelum, Pakistan
–Prof. M. Ashfaq Bukhari—B.Z. University, Multan, Pakistan
–Prof. Pål Davidsen—University of Bergen, Norway
–Prof. Mike Spector—The University of Bergen, Norway (now at the University of North Texas, USA)
–Prof. Yaman Barlas—The University of Bergen, Norway (now at the Bogazici University, Turkey)

–Prof. James Ang—National University of Singapore, Singapore
–Prof. Khalid Saeed—Worcester Polytechnic Institute (WPI), Worcester, USA
–Prof. Markus Schwaninger—University of St. Gallen, Switzerland

Hassan Qudrat-Ullah

Preface

New technologies have transformed the teaching and learning process. Academic institutions around the world are in the process of adopting technology in teaching, professional development, and curriculum development. The adoption and use of instructional technology play an important role in shaping the future of higher educational institutions (HEIs). The growth of educational tools such as learning management systems (LMSs) in recent years has inspired HEIs all over the world to redefine their teaching and learning processes. Consequently, HEIs are making huge investments in infrastructure, equipment, technology, and professional development programs to improve their educational effectiveness. However, these decisions regarding huge investments in HEIs are generally made without considering the variables and cultural context that affect the actual users of the technology. A lack of understanding of these influential variables results in the implementation failure due to users' unwillingness to accept new technology and hence the new technology does not meet the anticipated benefits for these institutions.

Most of the technology adoption models were developed and tested in Western cultures. The technology adoption models such as the revised 'Unified Theory of Acceptance and Use of Technology' (UTAUT2) do not address cultural variables and lack of cross-cultural study in non-Western countries. Therefore, it is very important to achieve a better knowledge of the impact of culture and other variables on the adoption of LMS. Thus, the purpose of the study in this book is to employ the UTAUT2 model as a framework for determining behavioral intention linked with the adoption and use of an LMS among instructors at HEIs. Furthermore, this study extends the UTAUT2 model in the context of 'Hofstede's cultural dimensions' and 'technology awareness' as the moderators of technology adoption. One of the prime incentives for this research is to explore the viability of the UTAUT2 model in non-Western countries, such as the Middle East, and to suggest some ways institutions can improve the adoption of LMS among instructors of these higher educational institutions. Hence, this study offers the administrators of HEIs with the facility to recognize the variables that influence the instructors' adoption of LMS and to incorporate these influential variables into the planning, investment, and implementation phases for effective adoption and use of LMS.

To meet the aim of the research, this book contains nine chapters. The first chapter presents an overview of this research. In Chap. 2, we discuss the adoption of new technology in general. Chapter 3 presents the importance of LMS, the historical perspective of LMS, types of LMS and its users, and key features of LMS. Chapter 4 discusses the adoption of new technology in the cultural context of Saudi Arabia. Chapter 5 compares different technology adoption models, and then builds a model on the modified unified theory of acceptance and use of technology (UTAUT2). Chapter 6 explains the chosen methodology by emphasizing the important features, strengths, and weaknesses of various approaches and methods. Chapter 7 explains the process of data preparation, results, and preliminary data analysis and provides empirical evidence regarding instructors' adoption of LMS in HEIs in the context of the Middle East in general and Saudi Arabia in particular. Chapter 8 deals with a broader discussion on the analysis of the results of the quantitative and suggest a holistic strategy formulation model. Chapter 9 presents a summary of this book project. This chapter provides an overview of the key elements of research that was undertaken in this project and presents with the insights toward LMS adoption and implications for theory and practice regarding the use and adoption of the learning management system.

Dhahran, Saudi Arabia   Rashid A. Khan
April 2020   Hassan Qudrat-Ullah

Acknowledgments

The first author (Rashid) would like to show his deep thanks to Dr. Carl Adams (University of Portsmouth, UK), who provided him the best support, experience, and knowledge. Rashid would also like to thank Mr. Qamar Iftikhar and Dr. Tahir Iqbal (National University of Sciences and Technology-NUST), Pakistan, for helping him with data analysis and interpretation. Hassan would like to thank Anam Qudrat for her timely review and edits of several pre-final chapters of this book. We would also like to acknowledge the support from Rehaneh Majidi (Springer).

Finally, the authors would like to acknowledge the financial support provided by the Deanship of Research at King Fahd University of Petroleum and Minerals for funding this work through project No. BW192-MGTMKT-92.

Dhahran, Saudi Arabia

Rashid A. Khan
Hassan Qudrat-Ullah

Contents

Adoption of LMS in the Cultural Context of Higher Educational Institutions of the Middle East

1.1 Introduction

Today, information and communication technology impacts almost all facets of life at the individual, organizational, and social levels. Academic institutions around the world are in the process of adopting technology in teaching, professional development, and curriculum development (Usluel, Askar, & Bas, 2008). The use of a learning management system (LMS) is growing in higher education. It has become an essential tool for instructors and students in their teaching and learning process (Alghamdi, 2016; Nagy, 2016). LMS is now equally popular not only in the educational sector but also in governmental organizations and business sectors for the training of individuals in the business sector. It helps corporate designers and educators to plan, organize, and deliver online courses in an effective way (Jafari, McGee, & Carmean, 2006).

Numerous models and theories have been presented to examine the variables that influence the adoption of new technologies (Baptista & Oliveira, 2015; Rodrigues, Sarabdeen, & Balasubramanian, 2016), in particular LMSs in education. These variables have been identified by a variety of information system theories and models, such as the theory of reasoned actions, technology acceptance model, and diffusion of innovation. A widely used technology adoption model, the unified theory of acceptance and use of technology (UTAUT), was introduced by Venkatesh et al. (2003), which united eight popular models of technology acceptance. Later, Venkatesh et al. (2012) revised the model UTAUT and included new constructs, namely, price value, hedonic motivation, and habit in the context of mobile phone consumer research. The new model is known as UTAUT2. Most of the technology adoption models were established and tested in developed countries. However, there is little research on the adoption and use of technology in non-Western countries (Al-Gahtani, Hubona, & Wang, 2007).

It was pointed out by Hofstede (1980) and other researchers (such as Baptista & Oliveira, 2015; Hew, Latifah, & Abdul, 2016) that culture has a remarkable influence on individuals in the adoption of technology. Hofstede (1980) provided cultural dimensions as theoretical grounds for exploring the acceptance of the technology. He described the cultural dimensions as power distance, masculinity, individualism, and uncertainty avoidance. Later, two more dimensions (indulgence and monumentalism) were also included. This book will provide empirical evidence to the validity of this extended model. Building on this work, we present a new cultural adoption model in the context of the Middle East. Here, in this chapter, we first describe the genesis of this book by presenting the context, problem statement, and the critical research questions. Finally, we present a content overview of this book.

1.1.1 Why This Book Matter?

New technologies have transformed the teaching and learning process (Henriksen, Mishra, & Fisser, 2016). The adoption and use of instructional technology play an important role in shaping the future of HEIs (Mosa, Naz'ri bin Mahrin, & Ibrrahim, 2016). The growth of educational tools such as LMSs in recent years (Siemens, 2015) has inspired HEIs all over the world to redefine their teaching and learning processes. Consequently, universities are making investments in equipment, technology, infrastructure, and professional development programs to improve their educational effectiveness (Tosunta, Karada, & Orhan, 2015). However, these decisions are generally made without considering the variables and dimensions that affect the actual users of the technology, that is, the instructors and students (Oblinger & Oblinger, 2005). A lack of understanding of the influential variables results in the implementation failure due to users' unwillingness to adopt new technology (Dillon & Morris, 1996; Mugo, Njagi, Chemwei, & Motanya, 2017) and hence these technological investments do not meet the anticipated benefits for these institutions. The institutions are lagging in the adoption of

© Springer Nature Switzerland AG 2021
R. A. Khan and H. Qudrat-Ullah, *Adoption of LMS in Higher Educational Institutions of the Middle East*, Advances in Science, Technology & Innovation, https://doi.org/10.1007/978-3-030-50112-9_1

available technology (Khan & Adams, 2016), and the research has found serious obstacles to fully integrate technology into educational processes (Cuban, Kirkpatrick, & C.P, 2001).

An exploratory study was conducted at King Fahd University of Petroleum and Minerals (KFUPM), Saudi Arabia to identify if there is a lack of LMS adoption in the institution. After the exploratory study, it was discovered that LMS technology has been made available by the academic institution but it is not being used to its full potential. A similar problem was identified by Cheng, Wang, Moormann, Olaniran, and Chen (2012), Dutton, Cheong, and Park (2004), and Khan and Adams (2016). The built-in functionalities and features of LMS systems to improve teaching and learning services are also underutilized (Sharma et al., 2011). Therefore, it is essential to investigate the determinants that influence the instructors' behavioral intentions to use the LMS at SHEIs.

This book addresses the gaps that have been identified through the literature review on the adoption of technology, variables related to technology adoption, models of technology acceptance, and cultural theories. Most technology acceptance models were established and tested in Western cultures (Al-Gahtani et al., 2007). However, a few published studies explored the adequacy of the models in non-Western cultures, especially Saudi Arabia. It would be naïve to assume that such a technology adoption model can be equally applicable in all cultural settings, especially in developing nations such as Saudi Arabia (Al-Gahtani et al., 2007). It is a well-recognized fact that cultural characteristics play a key role in technology adoption, yet cultural variables are ignored in most technology adoption models (Lin, 2014). Many researchers argue that cultural variables need to be incorporated in technology adoption models (Baptista & Oliveira, 2015; Lu & Lin, 2012) because information technology used by the people is impacted by cultural values (Im, Hong, & Kang, 2011). The original UTAUT2 model by Venkatesh et al. (2012) does not talk about cultural variables and lacks cultural awareness in the non-Western countries. Furthermore, the original UTAUT2 model was established and validated in the context of mobile phone consumer research. The literature shows that it has not been extensively adopted in educational settings to test the acceptance and use of LMS. Therefore, it is very important to achieve a better knowledge of the impact of culture and other variables on the adoption of LMS. Thus, the purpose of this research is to employ the extended version of 'unified theory of acceptance and use of technology' (i.e., UTAUT2) as a framework for determining behavioral intention linked with the adoption and use of an LMS among instructors at HEIs. The study in this book also attempts to determine the validity of the UTAUT2 model in non-Western cultures. Thus, the study in this book extends the UTAUT2 model with

Hofstede's (1980) cultural dimensions and technology awareness (TA) as moderators of the model. With the extension of the UTAUT2 model, it forms a new theoretical model that could help understand user behavior associated with the adoption of LMS in the cultural context of higher educational institutions (HEIs). One of the prime incentives for this research is to explore the viability of the UTAUT2 model in non-Western countries, such as Saudi Arabia, and to suggest some of the ways the institutions can improve the adoption of LMS among instructors of these higher educational institutions.

1.1.2 Research Objective and Research Questions (RQs)

Before defining RQs, it is imperative to understand the definitions of independent and dependent variables used in this research.

- **Performance Expectancy (PE)**: Venkatesh et al. (2003) defined performance expectancy as "the degree to which an individual believes that using the system will help him or her to attain gains in job performance" (p. 447).
- **Effort Expectancy (EE)**: Venkatesh et al. (2003) defined that effort expectancy is "the degree of ease associated with the use of the system" (p. 450).
- **Social Influence (SI)**: Social influence (SI) includes the social pressure exercised on a person by the beliefs of other individuals or groups. The social influence is "the degree to which an individual perceives that important others believe he or she should use the new system" (p. 451).
- **Facilitating Conditions (FC)**: Venkatesh et al. (2003) defined that facilitating conditions are "the degree to which an individual believes that an organizational and technical infrastructure exists to support the use of the system" (p. 453).
- **Hedonic Motivation (HM)**: Venkatesh et al. (2012) defined hedonic motivation as "the fun or pleasure derived from using a technology" (p. 161).
- **Habit (H)** is the automatic behavior that enables learning on how to use the technology. In other words, habit is the automaticity of behavior associated with the use of technology over time. Venkatesh et al. (2012) cited Limayem, Hirt, and Cheung (2007) that "habit is the extent to which people tend to perform behaviors automatically because of learning" (p. 161).
- **Use Behavior (UB)** is the actual use of the technology (Venkatesh et al. 2012).
- **Behavioral Intention (BI)**: According to the theory of reasoned action by Fishbein and Ajzen (1975) and theory

of planned behavior by Ajzen (1985), the greatest determinant of action is the 'intention', which is the user's willingness to perform a particular action.

The study in this book intends to answer the following central research question: *To what extent do independent variables (such as EE, PE, SI, FC, HM, and H) and moderating variables of the proposed model influence instructors' behavioral intentions to use LMS in HEIs?* Thus, this research identifies the following research sub-questions:

- **RQ 1**: To what extent (if any) is behavioral intention (BI) a predictor of use behavior (UB) of LMSs at HEIs?
- **RQ 2**: To what extent (if any) do independent variables (EE, PE, SI, FC, HM, and H) affect instructors' behavioral intentions to adopt an LMS at HEIs?
- **RQ 3**: Which out of the six independent variables (EE, PE, SI, FC, HM, and H) delivers the most significant contribution to instructors' behavioral intentions to adopt an LMS at HEIs?
- **RQ 4**: To what extent (if any), do moderate variables moderate the relationship between the dependent and independent variables?

1.2 The Approach of This Research Project

In this research, the explanatory technique is selected because in the context of higher educational institutions, the variables (of UTAUT2) for the research were largely well known in the literature, and statistical techniques (SPSS/Amos) are used to identify the significant variables. The quantitative data were collected via a survey questionnaire. This research used structural equation modeling (SEM) to assess the adequacy of the conceptual model and the measurement model by the SPSS/Amos program. In this research project, we used the exploratory factor analysis (EFA), followed by confirmatory factor analysis (CFA) that validates the measurement model in the proposed research model.

1.3 Contribution and Significance of the Research

The research accomplished and reported in this book contributes to reducing the gap in scholarly research regarding instructors' perceptions of the adoption of LMS at HEIs. This research identifies the significant variables that influence instructors' behavioral intentions to adopt an LMS in the cultural context of HEIs. Hence, this research offers the administrators of HEIs with the facility to recognize the

variables that influence the instructors' adoption of LMS and to incorporate these influential variables into the planning, investment, and implementation phases for effective adoption and use of LMS.

To capture the ignored variables, this research extends the UTAUT2 model into three dimensions. The suggested amalgamated model attempts to address the limitations of the original UTAUT2 model by integrating new constructs in the context of HEIs. The inclusion of additional variables makes it the first study of its kind applied to the UTAUT2 model in higher educational institutions (HEIs). This research also confirms the viability of the UTAUT2 model in non-Western countries such as Saudi Arabia. This research adds to the body of knowledge on LMS by proposing a theoretical model that integrates Hofstede's (1980) cultural dimensions with the UTAUT2 model (Baptista & Oliveira, 2015), providing novel insights into the cultural context of HEIs.

In this book, we adopt the instruments established by Venkatesh et al. (2012) and Hofstede's (1980) value survey module (VSM) with some modifications suitable for educational settings. The extended instrument offers new means for further research in the adoption of LMS in non-Western countries. By using the results of the research presented in this book, the administrators of HEIs may encourage and convince instructors and students of how easy it is to use and how useful this new technology is for them, thereby encouraging an increase in the use of the LMS in their teaching and learning processes.

1.4 The Organization of This Book

To meet the objective of advancing the use and appreciation of LMS in higher education in the context of the Middle East, this book contains nine chapters. This chapter presents an overview of the genesis of this book through the description of the background, statement of the problem, gaps in the literature, aims and objectives, approach, and significance of this research. In Chap. 2, we discuss the adoption of new technology in general. Chapter 3 presents the importance of LMS, the historical perspective of LMS, types of LMS and its users, and key features of LMS. Chapter 4 discusses the adoption of new technology in the cultural context of Saudi Arabia. Chapter 5 compares different technology adoption models, and then builds a model on the modified unified theory of acceptance and use of technology (UTAUT2). This research presented in this book extends the UTAUT2 model with 'technology awareness' and 'Hofstede's cultural dimensions' as moderating variables of the UTAUT2 model. In Chap. 6, we explain the chosen methodology by emphasizing the important features, strengths, and weaknesses of various approaches and methods. It includes the research design, survey instruments, data collection, and statistical techniques for analyses. It also describes the procedure followed and describes the motives behind the specific techniques, methods, and approaches of data collection. Finally, Chap. 6 discusses the reliability, validity, and associated issues. Chapter 7 explains the process of data preparation, results, and preliminary data analysis. The prime focus of this chapter is the appropriateness of the data obtained concerning data analysis. This chapter presents the preliminary data results and the statistical methods applied in data analysis. In this chapter, the personal profile of the participants and the descriptive data analysis applied are discussed. The descriptive data analysis of the core variables (PE, EE, SI, FC, HM, and H) and moderating variables (experience, age, and cultural dimensions) are discussed. In the last section, the reliability, correlation, factor analysis, and regression analysis are discussed. Hence, this chapter provides empirical evidence regarding instructors' adoption of LMS in HEIs in the context of the Middle East in general and Saudi Arabia in particular. Chapter 8 deals with a broad discussion on the analysis of the results of the quantitative and suggest a holistic strategy formulation model. This chapter deals with a broad discussion on the analysis of the results of the quantitative data and suggests a holistic strategy formulation model. The key research questions have been addressed in this chapter. Finally, Chap. 9 presents a summary of this book project. This chapter provides an overview of the key elements of research that was undertaken in this project. The contributions and insights toward LMS adoption, implications for theory and practice regarding the use and adoption of LMS, the limitations of the research presented here in this book are discussed and the recommendations for future research about wider acceptance and utility of learning management systems across the work are made.

1.5 Summary

The adoption and use of instructional technology play an important role in shaping the future of HEIs. The use of a learning management system (LMS) is growing in higher education. It has become an essential tool for instructors and students in their teaching and learning process. Various models and theories have been offered to investigate the factors that influence the adoption of LMSs in the educational sector. Most technology adoption models were developed and tested in Western cultures. Therefore, it is very important to achieve a better knowledge of the impact of cultural variables on the adoption of LMS. Thus, the

purpose of the study in this book is to employ the extended version of 'unified theory of acceptance and use of technology' (i.e., UTAUT2) as a framework for determining behavioral intention linked with the adoption and use of an LMS among instructors at HEIs. This book provides empirical evidence to the validity of the extended UTAUT2 model in the cultural context of higher educational institutions. With the extension of the UTAUT2 model, it forms a new theoretical model that could help understand user behavior associated with the adoption of LMS in the cultural context of higher educational institutions (HEIs).

Hence, this chapter first presents the introduction of the book by presenting the context, problem statement, and critical research questions. Finally, this chapter presents a summary of all chapters of this book.

References

Ajzen, I. (1985). From intentions to actions: A theory of planned behavior. In J. Kuhl & J. Beckman (Eds.). *Action–control: From cognition to behavior* (pp. 11–39), Heidelberg: Springer.

Al-Gahtani, S. S., Hubona, G. S., & Wang, J. (2007). Information technology (IT) in Saudi Arabia: Culture and the acceptance and use of IT. *Information & Management, 44*(8), 681–691. https://doi.org/10.1016/j.im.2007.09.002.

Alghamdi, S. R. (2016). Use and attitude towards Learning Management Systems (LMS) in Saudi Arabian universities. *Eurasia Journal of Mathematics, Science & Technology Education, 12*(9), 2309–2330. https://doi.org/10.12973/eurasia.2016.1281a.

Baptista, G., & Oliveira, T. (2015). Understanding mobile banking: The unified theory of acceptance and use of technology combined with cultural moderators. *Computers in Human Behavior, 50*, 418–430. https://doi.org/10.1016/j.chb.2015.04.024.

Cheng, B., Wang, M., Moormann, J., Olaniran, B. A., & Chen, N.-S. (2012). The effects of organizational learning environment factors on e-learning acceptance. *Computers & Education, 58*(3), 885–899. https://doi.org/10.1016/j.compedu.2011.10.014.

Cuban, L., Kirkpatrick, H., & P, C. (2001). High access and low use of technologies in high school classrooms: Explaining an apparent paradox. *American Educational Research Journal, 38*(4), 813–814.

Dillon, A., & Morris, M. G. (1996). Users acceptance of information technology: theories and models. *Annual Review of Information Science and Technology, 31*, 3–32.

Dutton, H., Cheong, P., & Park, N. (2004). The social shaping of a virtual learning environment. *Electronic Journal of E-Learning, 2* (2), 1–12. Retrieved from http://www.inf.ufes.br/~cvnascimento/artigos/issue1-art3-dutton-cheong-park.pdf.

Fishbein, M., & Ajzen, I. (1975). *Belief, attitude, intention, and behavior: An introduction to theory and research*. Don Mills, Ontario: Addison-Wesley Publishing Company.

Henriksen, D., Mishra, P., & Fisser, P. (2016). Infusing creativity and technology in 21st century education: a systemic view for change. *Journal of Educational Technology & Society Educational Technology & Society, 19*(193), 27–37. http://about.jstor.org/terms.

Hew, T., Latifah, S., & Abdul, S. (2016). Computers & education understanding cloud-based VLE from the SDT and CET perspectives: Development and validation of a measurement instrument. *Computers & Education, 101,* 132–149. https://doi.org/10.1016/j.compedu.2016.06.004.

Hofstede, G. (1980). Culture and organizations. *International Studies of Management & Organization, 10*(4), 15–41.

Im, I., Hong, S., & Kang, M. S. (2011). An international comparison of technology adoption. *Information & Management, 48*(1), 1–8. https://doi.org/10.1016/j.im.2010.09.001.

Jafari, A., McGee, P., & Carmean, C. (2006). Managing courses defining learning: What faculty, students, and administrators want. *Educause Review, 4*(4), 50–51.

Khan, R. A., & Adams, C. (2016). Adoption of learning management systems in Saudi higher education context : Study at King Fahd University of Petroleum and Minerals & Dammam Community College. In *Society for Information Technology & Teacher Education International Conference* (Vol. 2016. No. 1, pp. 2909–2916). Barbera 2004.

Limayem, M., Hirt, S. G., & Cheung, C. M. (2007). How habit limits the predictive power of intention: The case of information systems continuance. *MIS Quarterly, 31*(4), 705–737.

Lin, H.-C. (2014). An investigation of the effects of cultural differences on physicians' perceptions of information technology acceptance as they relate to knowledge management systems. *Computers in Human Behavior, 38*(April), 368–380. https://doi.org/10.1016/j.chb.2014.05.001.

Lu, H.-K., & Lin, P.-C. (2012). Toward an extended behavioral intention model for e-learning: Using learning and teaching styles as individual differences. In *2012 2nd International Conference on Consumer Electronics, Communications and Networks (CECNet)* (pp. 3673–3676). https://doi.org/10.1109/CECNet.2012.6202261.

Mosa, A. A., Naz'ri bin Mahrin, M., & Ibrrahim, R. (2016). Technological aspects of e-learning readiness in higher education: A review of the literature. *Computer and Information Science, 9*(1), 113. https://doi.org/10.5539/cis.v9n1p113.

Mugo, D., Njagi, K., Chemwei, B., & Motanya, J. (2017). The technology acceptance model (TAM) and its application to the utilization of mobile learning technologies. *British Journal of Mathematics & Computer Science, 20*(4), 1–8. https://doi.org/10.9734/BJMCS/2017/29015.

Nagy, J. T. (2016). Using learning management systems in business and economics studies in Hungarian higher education. *Education and Information Technologies, 21*(4), 897–917. https://doi.org/10.1007/s10639-014-9360-6.

Oblinger, D., & Oblinger, J. (2005). Is it age or IT: First steps toward understanding the net generation. *Educating the Net Generation, Chapter 2*(2), 2.1–2.20.

Rodrigues, G., Sarabdeen, J., & Balasubramanian, S. (2016). Factors that influence consumer adoption of e-government services in the UAE: A UTAUT model perspective. *Journal of Internet Commerce, 15*(1), 18–39. https://doi.org/10.1080/15332861.2015.1121460.

Sharma, S. A. T., Paul, A., Gillies, D., Conway, C., Nesbitt, S., Ripstein, I. R. A., Simon, I., & Mcconnell, K. (2011). Learning/Curriculum Management Systems (LCMS): Emergence of a New Wave in Medical Education. *Learning, 11*(13).

Siemens, G. (2015). Learning analytics: The emergence of a discipline. *American Behavioral Scientist, 57*(10), 1380–1400. https://doi.org/10.1177/0002764213498851.

Tosunta, B., Karada, E., & Orhan, S. (2015). The factors affecting acceptance and use of interactive whiteboard within the scope of FATIH project: A structural equation model based on the unified theory of acceptance and use of technology. *Computers & Education, 81,* 169–178. https://doi.org/10.1016/j.compedu.2014.10.009.

Usluel, Y. K., Askar, P., & Bas, T. (2008). A structural equation model for ICT usage in higher education. *Educational Technology & Society, 11*(2), 262–273.

Venkatesh, V., Thong, J., & Xu, X. (2012). Consumer acceptance and use of information technology: Extending the unified theory. *MIS Quarterly, 36*(1), 157–178.

Venkatesh, V., Morris, M. G., Davis, G. B., & Davis, F. D. (2003). User acceptance of information technology: Toward a unified view. *MIS Quarterly, 27*(3), 425–478.

2.1 Introduction

Technology pervades almost all areas in society (Scherer, Siddiq, & Tondeur, 2019). To improve the quality of education and to enhance the professional productivity of the instructors, the latest technologies are required to be adopted and integrated into the educational system (Tomei, 2005). Many colleges and universities are restructuring their existing degree programs and integrating new technology into teaching to enhance their educational process (Buabeng-Andoh, 2012). Thus, this chapter discusses the adoption of new technology in the cultural context of higher educational institutions.

2.2 What Is Technology Adoption

The meaning of 'adoption' is to make full use of technology and 'rejection' is a judgment not to accept the technology (Rogers, 2003). Adoption is a process of technology awareness and then making full use of it (Wong, 2016). It is well documented by researchers that the decision of an individual regarding the adoption of technology is not a sudden process (Straub, 2009) but adoption occurs with the passage time and involves a series of many actions (Rogers, 2003). However, the adoption of technology involves an individual's readiness to use a technology for which it was designed (Wong, 2016). According to the Rogers' definition (1983), 'acceptance' is a process during which, based on his/her primary knowledge, an individual develops an attitude toward that innovation and intention to adopt that innovation and insists on this decision (Keramati, Sharif, Azad, & Soofifard, 2012). User acceptance of technology may be described as a person's psychological condition regarding his/her intention to use a technology (Dillon & Morris, 1996). The behavioral intention has been extensively used in all the prior literature (Keramati et al., 2012). Most of the studies in technology adoption are based on the behavioral intention of an individual.

Although there is a difference between technology adoption and technology acceptance (Wong, 2016), this study will use the adoption and acceptance of technology interchangeably. Information and communication technology (ICT) in educational institutions is generally known as educational technology (Wong, 2016). Educational technology has a broader meaning and refers to the use of technology for learning and teaching purposes by the students or by the instructors (Wong, 2016). Today ICT influences almost all facets of life at the individual, organizational, and social levels. The use of technology has become a vital prerequisite for the development of a knowledge-based economy (Murshitha & Wickramarachchi, 2016; Zanjani, Edwards, Nykvist, & Geva, 2016). Institutions around the world are in the process of adopting technology in teaching, professional development, and curriculum development (Pulkkinen, 2007; Usluel, Askar, & Bas, 2008). The potential benefits of technology in learning and teaching have received substantial attraction in recent years (Wong, 2016). The students can now have access to unlimited information, have more control of their education, and can use different modes of communication and learning. The growth of telecommunication and digital information technologies has had a deeper effect on individuals, society, business, and education. Bitter and Legacy (2008) stated that high-speed Internet allows students and teachers have access to online materials around the clock. The arrival of the Internet has modified the role of instructors by providing comprehensive information to students in the classroom. The growing electronic learning (e-learning) environment in the teaching environment is an unavoidable phenomenon. Mobile learning (m-learning) has more flexibility and can be independent of place and time; the institutions can blend a variety of approaches for on-campus and off-campus learners. According to Georgiev, Georgieva, and Smrikarov (2004) m-learning and e-learning environments provide the most flexibility in teaching and learning.

© Springer Nature Switzerland AG 2021

R. A. Khan and H. Qudrat-Ullah, *Adoption of LMS in Higher Educational Institutions of the Middle East*, Advances in Science, Technology & Innovation, https://doi.org/10.1007/978-3-030-50112-9_2

2.3 Evolution of Technology in Education

Gilbert and Green (1995) stated in their book '*Information Technology: A Road to the Future*' that technology has evolved from a slide rule to networked computers to communicate instantly with the world. The technology has deep roots and impacts in the educational field. About 2500 years ago, memorization was the way to retain knowledge and skills, whereas communication was generally oral. Technology has been used in teaching and learning for many years and has a positive influence on learning and teaching (Wong, 2016). The use of technology such as computers to support educationalists started in the late 1950s and still, it is evolving (Tamim, Bernard, Borokhovski, Abrami, & Schmid, 2011). Mainframe computers entered into the educational sector in the 1950s. Computers at that time were slow, not flexible. Computer-aided instruction was started in the 1960s, but it was still linear and was not flexible (VanDusen, 1997). The microprocessor was a key evolution and it became the heart of all innovations (Stark & Lattuca, 1997). In the 1970s, the IBM personal computer and Apple-II were introduced to the market, and the desktop computer eventually turned out to be very popular (Norman, 1998). The development of personal computers (PC) placed control in the hands of consumers and brought a digital revolution in computing, business, gaming, and education. The early 1990s desktop computers became affordable and during the latter part of the 1990s, the Internet and the facsimile machine became popular among the people (Stark & Lattuca, 1997). Laptop computers and smartphones with Wi-Fi technologies provided connectivity to local wireless networks. The further advance in technology is the connectivity of smart devices, where users are connected using a built-in broadband card which keeps them connected to Internet service providers' networks. Mobile communication technology through smart mobile phones is the new frontier for educational experts for its potential of enhancing the teaching and learning process anywhere and everywhere (Lan, Tsai, Yang, & Hung, 2012). The Internet was one of the major steps in the development of the digital age that had a significant influence on the economy, society, business, education, and global communications. The development of the World Wide Web (WWW) expedited the exponential growth of the information revolution and modernized the transfer of information by making it available to everyone, anytime, and everywhere (De Freitas & Levene, 2003). Today, the entire world has become one global village reachable to everyone from anywhere, where shopping, communications, and information are just a mouse click away. This incredibly fast evolution of the Internet from 1995 to the present time is depicted in Table 2.1.

Internet technology continues to grow day-by-day making a global village a reality (Miniwatts Marketing Group, 2016). Table 2.1 shows the extraordinary growth of the Internet from 1995 until 2019. It is clear from the table that the worldwide number of Internet users was 16 million in December 1995 (0.4% of the world population). The number of Internet users was 1971 million in 2010 (28.8% of the world population), 3366 million in December 2015 (46.4% of the world population), 4156 million in 2017 (54.4% the

Table 2.1 History of the internet from December 1995 to June 2019 and the number of users

Date	Number of users	% World population	Date	Number of users	% World population
December 1995	16 millions	0.40	December 2008	1574 millions	23.50
December 1996	36 millions	0.90	December 2009	1802 millions	26.60
December 1997	70 millions	1.70	September 2010	1971 millions	28.80
December 1998	147 millions	3.60	December 2011	2267 millions	32.70
December 1999	248 millions	4.10	December 2012	2497 millions	35.70
December 2000	361 millions	5.80	December 2013	2802 millions	39.00
August 2001	513 millions	8.60	December 2014	3079 millions	42.40
September 2002	587 millions	9.40	December 2015	3366 millions	46.40
December 2003	719 millions	11.10	December 2016	3696 millions	49.50
December 2004	817 millions	12.70	December 2017	4156 millions	54.40
December 2005	1018 millions	15.70	December 2018	4313 millions	55.60
December 2006	1093 millions	16.70	June 2019	4422 millions	57.30
December 2007	1319 millions	20.00	January 2020	4574 millions	58.7

Source https://www.internetworldstats.com/emarketing.htm

world population), 4313 million in 2018 (55.6 8% of the world population), and 4422 million (57.3% of the world population).

2.4 Evolution of Learning Paradigms

Teaching and learning have been carried out through the traditional face-to-face (F2F), time and place-restricted mode, and it was considered the backbone of the educational system. The technology-assisted teaching and learning model has migrated from a traditional model to a distance model. The system of education is in the exploration phase of this new learning system and is transforming from the traditional model to d-learning, e-learning, and m-learning models.

2.4.1 Distance Learning (D-Learning)

The concept of distance learning is not new and was started around a century ago as correspondence courses in Europe (Lowell, Harris, Lowell, & Editors, 2018; Valentine, 2002). This method of correspondence converted into another form when educational television and radio achieved popularity. Distance learning is an education process in which the learners and the educators are separated by time or location, or both. Education is provided to remote locations asynchronously or synchronously using audiotapes, videotapes, CD-ROM, online, or video conferencing (Nagy, 2016). Greenberg (1998) describes the process of education, free from the need of traveling that makes the use of the Internet to connect the students and to interact with them. Valentine (2002) defines that distance learning *"is the separation between the students and instructor by place, but not essentially by the time"*. Through the 1970s and 1980s, telecourses remained popular as part of the distance learning movement. Television was introduced with the distance learning movement. One of the reasons for the adoption of this mode of instruction is the increased pressure and competition on institutions to generate revenue, control costs, and meet industry requirements (Valentine, 2002). Despite the problems reported with the quality of equipment and teaching, the literature is well documented by the increased demands of distance education (Ferguson & Wijekumar, 2000). The information technology and internet communication opened the doors for electronic modes of reaching the students and eventually shifting to e-learning teaching. Many universities are offering their courses via e-learning or online learning.

2.4.2 Electronic Learning (E-Learning)

The information revolution, also known as electronic revolution (e-revolution), has impacted at individual, social, economic, business, and government levels. The introduction of an e-environment, such as e-learning, e-commerce, e-mail, e-banking, and e-government, brought new opportunities and challenges for individuals, societies, businesses, educational institutions, and governments (Nagy, 2016). e-learning is a learning using media and digital electronic tools (Pinkwart, Hoppe, Milrad, & Perez, 2003). Rosenberg (2001) defines e-learning as a type of education that depends on network and Internet technology. The demand for e-learning-based courses is increasing in universities (Murshitha & Wickramarachchi, 2016). An e-revolution made it possible for educational sectors to accommodate distance learners (Collis & Moonen, 2002) and the concept of online education received rapid popularity (Matheos, Daniel & McCalla, 2005). There are now various 'virtual-only' universities that are offering e-learning courses. Examples of such universities include the World Lecture Hall of the University of Texas, the British Open University, the Globewide Network Academy in Denmark, and the Athena University. Many prestigious universities have joined together and developed non-profit alliances to create d-learning programs. There are *advantages* and *disadvantages* of e-learning. The advantages of e-learning include: e-learning is not limited by physical location. It is quicker because the students may skip the contents they previously knew. For instance, more than 96% of the courses are offered as 'online' in most colleges and universities (Sharma et al., 2011). e-learning provides the opportunity for students to customize their materials according to their requirements. In this way, the students achieve control over the learning process and a better understanding that make the learning process more effective. It is more flexible, self-paced, and cheaper to deliver. The online material can be updated quickly and easily because it is done on the server-side. The students can start training at any time and from anywhere when they need. It is beneficial for educators to manage large classes. The communication and interaction between instructors and students is another advantage of e-learning through the use of instant messaging, discussion boards, chat rooms, and email. Cantoni, Cellario, and Porta (2003) describe the following disadvantages: It may cost more initially to develop the new system. e-learning technology might be threatening, confusing, or frustrating. The key disadvantage of e-learning is lacking part of the F2F interaction and informal social interaction. Developing new courses for new e-learning to achieve new skills also requires money, time, and training. New technology often causes frustration due to a lack of training, especially during its adoption process. With appropriate and careful design, most of the shortcomings can be overcome (Cantoni et al., 2003).

2.4.3 Mobile Learning (M-Learning)

M-learning is an extension of e-learning using portable and mobile learning devices (Doneva, Nikolaj, & Totkov, 2006). m-learning is a new phase of e-learning with learning capability from everywhere at any time using mobile devices (Georgiev et al., 2004). The point at which e-learning and mobile computing intersect, an anytime, anywhere learning experience is developed (Ismail, 2016; Kambourakis, Kontoni, & Sapounas, 2004). m-learning is a flexible learning paradigm of anywhere and anytime (Wagner, 2005). m-learning provides the learners with additional learning provision, a broader channel of communication, and flexibility of access (McConatha, Matt, & Lynch, 2008). New technologies linked with the m-revolution are influencing and shaping the boundaries of economy and knowledge domains (Wagner, 2005). M-technology has altered the use and scale of m-learning. Unlike a book, thousands of books can be stored on a palm device, interactive media can play in a palm device, and the m-device can virtually take you into the classroom. Smartphones, PDA, MP3 players, and GPS devices are examples of such devices. m-learning has altered the concept of the classroom by facilitating communication between students and instructors (Wentzel et al., 2005). Information technology is expanding the borders of teaching and learning into 'anywhere and anytime'. Mobile communication through smart mobile devices is not restricted by location or time and provides access to teaching and learning. The students now have more control over their literacy tools and educational needs. These new mobile technologies have shaped new opportunities as well as challenges for individuals, businesses, institutions, and governments (Gloria, 2016).

2.5 Difference Between D-Learning, E-Learning, and M-Learning

It is important to note the difference between d-learning and e-learning. The terms 'e-learning', 'web-based learning', 'distance learning', and 'online learning' are often used interchangeably (Littlefield, 2016). However, there is a

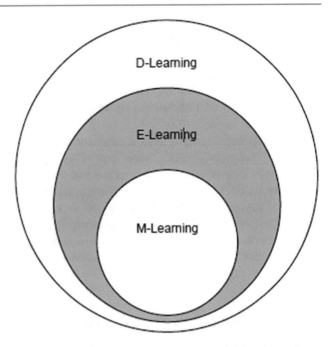

Fig. 2.1 M-learning as part of e-learning and d-learning. *Source* Georgiev et al. (2004)

difference between d-learning and e-learning. Georgiev et al. (2004) showed the relationship between m-learning, e-learning, and d-learning in Fig. 2.1. e-learning is any kind of learning that includes the use of technology to help the learner. The term just refers to the tools used. e-learning can happen right in the classroom in face-to-face learning. It can be used when teacher and student are separated too. d-learning is an education process in which the learners and the educators are separated by time or location, or both. Education is provided to remote locations asynchronously or synchronously using audiotapes, videotapes, CD-ROM, online, or video conferencing (Nagy, 2016).

As it is shown in Fig. 2.1, m-learning is considered to be a subset of e-learning, and e-learning is considered to be a subset of d-learning (Georgiev et al., 2004). Tick (2006) explained the relationship between m-learning, e-learning, and d-learning.

D-learning is gradually changing to e-learning due to rapid technological inventions. This relationship is illustrated in Fig. 2.2. The progression of m-technologies such as smartphones, tablet PCs, and personal digital assistants (PDA) has attracted the researchers and educators to ponder its pedagogical implications (McConatha et al., 2008).

Fig. 2.2 The interrelationship of m-learning, e-learning, and d-learning. *Source* Tick (2006)

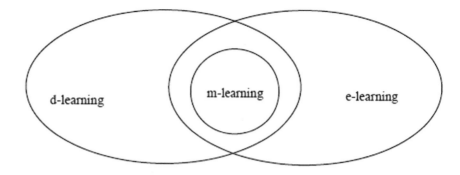

2.6 Future Direction of E- and M-Learning

A blended approach is a hybrid of both online learning and traditional F2F learning (Chen, Yong, & Yao 2016; Collis & Moonen, 2002). In other words, a 'blended' environment combines both traditional teaching methods of F2F classes and e-teaching methods (Mortera-Gutierrez, 2006). The use of e-technology does not mean removing traditional F2F contact from the educational culture but improving and supporting the teaching and learning process by using new technologies. LMS is the future direction of e-learning (Alghamdi, 2016; Weller, 2007). Wide acceptance of a VLE (or LMS) highlighted the importance of the e-learning approach. LMSs such as WebCT and Blackboard are the components of digital e-learning and widely used for both on-campus and off-campus students at major universities of the world (Keegan, 2002; Murshitha & Wickramarachchi, 2016; Nagy, 2016).

2.7 Summary

The adoption of new technology plays an important role in shaping the future of the educational sector of any country (Mosa, Naz'ri bin Mahrin, & Ibrahim, 2016). Thus, it is important to understand the concept of users' adoption of technology. The system of education is in the exploration phase of this new learning system and is transforming from the traditional model to distance learning (d-learning), electronic-learning (e-learning), and mobile learning (m-learning) models. Thus, this chapter describes the role of d-learning, e-learning, and m-learning as learning platforms. LMS has the potential of improving online learning and creating a fully functional virtual classroom. Hence, an LMS is the future direction of e-learning.

References

Alghamdi, S. R. (2016). Use and attitude towards Learning Management Systems (LMS) in Saudi Arabian universities. *Eurasia Journal of Mathematics, Science & Technology Education, 12*(9), 2309–2330. https://doi.org/10.12973/eurasia.2016.1281a.

Bitter, G. G., & Legacy, J. M. (2008). *Using technology in the classroom* (7th ed.). Pearson/Allyn and Bacon Publishers.

Buabeng-Andoh, C. (2012). Factors influencing teachers' adoption and integration of information and communication technology into teaching: A review of the literature Charles Buabeng-Andoh. *International Journal of Education and Development Using Information and Communication Technology (IJEDICT), 8*(1), 136–155.

Cantoni, V., Cellario, M., & Porta, M. (2003). Perspectives and challenges in elearning: Towards natural interaction paradigms. *Journal of Visual Languages and Computing, 15*, 333–345.

Chen, W. S., Yong, A., & Yao, T. (2016). An empirical evaluation of critical factors influencing learner satisfaction in blended learning : A pilot study. *4*(7), 1667–1671. https://doi.org/10.13189/ujer.2016.040719.

Collis, B., & Moonen, J. (2002). Flexible learning in a digital world. Open learning. *The Journal of Open and Distance Learning, 17*(3), 217–230.

De Freitas, S., & Levene, M. (2003). *Evaluating the development of wearable devices, personal data assistants and the use of other mobile devices in further and higher education institutions.* Retrieved from http://eprints.bbk.ac.uk/archive/00000176/01/defreitas1.pdf.

Dillon, A., & Morris, M. G. (1996). Users acceptance of information technology: Theories and models. *Annual Review of Information Science and Technology, 31*, 3–32.

Doneva, R., Nikolaj, K., & Totkov, G. (2006). Towards mobile university campuses. In *International Conference on Computer Systems and Technologies - CompSysTech'2006.*

Ferguson, L., & Wijekumar, K. (2000). Effective design and use of web-based distance learning environments. *Professional Safety, 45* (12), 28–33.

Georgiev, T., Georgieva, E., & Smrikarov, A. (2004). M-learning-a new stage of e-learning. *In International Conference on Computer Systems and Technologies-CompSysTech*, 28. Retrieved from http://scholar.google.com/scholar?hl=en&q=http://ldt.stanford.edu/~educ39106/articles/m-learning.pdf&btnG=&as_sdt=1,5&as_sdtp=#.

Gilbert, S. W., & Green, K. C. (1995). *Information technology: A road to the future? To promote academic justice and excellence series.*

Gloria, A. (2016). Influence of mobile learning training on pre-service social studies teachers' technology and mobile phone self-efficacies. *7*(2), 74–79.

Greenberg, G. (1998). Distance education technologies: Best practices for K-12 settings. *IEEE Technology and Society Magazine, 36–40.*

Ismail, A. (2016). The effective adoption of ICT-enabled services in educational institutions—key issues and policy implications. *Journal of Research in Business, Economics and Management (JRBEM), 5*(5), 717–728.

Kambourakis, G., Kontoni, D.-P., & Sapounas, I. (2004). Introducing attribute certificates to secure distributed e-learning or m-learning services. *The IASTED International Conference* (pp. 16–18). http://www.ice.upc.edu/butlleti/innsbruck/416-174.pd.

Keegan, D. (2002). The future of learning: From eLearning to mLearning. In *ZIFF*. Hagen, Germany: Institute for Research into Distance Education. Retrieved from http://deposit.fernuni-hagen.de/1920/1/ZP_119.pdf.

Keramati, A., Sharif, H. J., Azad, N., & Soofifard, R. (2012). Role of subjective norms and perceived behavioral control of tax payers in acceptance of e-tax payment system. *International Journal of E-Adoption, 4*(3), 1–14. https://doi.org/10.4018/jea.2012070101.

Lan, Y.-F., Tsai, P.-W., Yang, S.-H., & Hung, C.-L. (2012). Comparing the social knowledge construction behavioral patterns of problem-based online asynchronous discussion in e/m-learning environments. *Computers & Education, 59*(4), 1122–1135. https://doi.org/10.1016/j.compedu.2012.05.004.

Littlefield, J. (2016). *What's the difference between e-Learning and distance learning?* Retrieved from https://www.thoughtco.com/e-learning-vs-distance-learning-3973927.

Lowell, V. L., Harris, B. R., Lowell, V. L., & Editors, B. R. H. (2018). *Leading and Managing* (Issue October 2017).

Matheos, K., Daniel, B. K., & McCalla, G. L. (2005). Dimensions for blended learning technology: Learners' perspectives. *Journal of Learning Design, 1*(1), 56–76.

McConatha, D., Matt, P., & Lynch, M. J. (2008). Mobile learning in higher education: An empirical assessment of a new educational tool. *The Turkish Online Journal of Educational Technology, 7*(3).

Miniwatts Marketing Group. (2016). *Internet growth statistics. Internet World Stats.* Retrieved from http://www.internetworldstats.com/emarketing.htm.

Mortera-Gutierrez, F. (2006). Faculty best practices using blended learning in e-learning and face-to-face instruction. *International Journal on ELearning, 5*(3), 313–337.

Mosa, A. A., Naz'ri bin Mahrin, M., & Ibrrahim, R. (2016). Technological aspects of e-learning readiness in higher education: A review of the literature. *Computer and Information Science, 9*(1), 113. https://doi.org/10.5539/cis.v9n1p113.

Murshitha, S. M., & Wickramarachchi, A. P. R. (2016). A study of students' perspectives on the adoption of LMS at University of Kelaniya. *Journal of Management, 9*(1), 16. https://doi.org/10.4038/jm.v9i1.7562.

Nagy, J. T. (2016). Using learning management systems in business and economics studies in Hungarian higher education. *Education and Information Technologies, 21*(4), 897–917. https://doi.org/10.1007/s10639-014-9360-6.

Norman, D. A. (1998). The invisible computer. *Embedded Systems Programming, 3453*(July), 1996. Retrieved from http://draconis-syndicate.de/texts/essays/TheInvisibleComputer.pdf.

Pinkwart, N., Hoppe, H. U., Milrad, M., & Perez, J. (2003). Educational scenarios for the cooperative use of personal digital assistants. *Journal of Computer Assisted Learning, 19*(3), 383–391.

Pulkkinen, J. (2007). Cultural globalization and integration of ICT in education. In K. Kumpulainen (Ed.), *Educational technology: Opportunities and challenges* (pp. 13–23).

Rogers, E. M. (2003). *Diffusion of innovations* (5th ed). Free Press.

Rosenberg, M. J. (2001). *E-learning: Strategies for delivering knowledge in the digital age.* McGraw-Hill.

Scherer, R., Siddiq, F., & Tondeur, J. (2019). The technology acceptance model (TAM): A meta-analytic structural equation modeling approach to explaining teachers' adoption of digital technology in education. *Computers & Education, 128*(0317), 13–35. https://doi.org/10.1016/j.compedu.2018.09.009.

Sharma, S. A. T., Paul, A., Gillies, D., Conway, C., Nesbitt, S., Ripstein, I. R. A., Simon, I., & Mcconnell, K. (2011). Learning/Curriculum Management Systems (LCMS): Emergence of a new wave in medical education. *Learning, 11*(13).

Stark, J. S., & Lattuca, L. R. (1997). Shaping the college curriculum. In *Shaping the college curriculum.* Needham Heights, MA: Allyn & Bacon.

Straub, E. T. (2009). Understanding technology adoption: Theory and future directions for informal learning. *Review of Educational Research, 79*(2), 625–649.

Tamim, R. M., Bernard, R. M., Borokhovski, E., Abrami, P. C., & Schmid, R. F. (2011). What forty years of research says about the impact of technology on learning a second-order meta-analysis and validation study. *Review of Educational Research, 81*(1), 4–28.

Tick, A. (2006). A Web-based e-learning application of self study multimedia programme in military English. In *3rd Romanian-Hungarian Joint Symposium on Applied Computational Intelligence-SACI 2006.* Retrieved from http://www.bmf.hu/conferences/saci2006/Andrea_Tick.pdf.

Tomei, L. A. (2005). *Taxonomy for the technology domain.* Information Science Publishing.

Usluel, Y. K., Askar, P., & Bas, T. (2008). A structural equation model for ICT usage in higher education. *Educational Technology & Society, 11*(2), 262–273.

Valentine, D. (2002). Distance learning: Promises, problems, and possibilities. *Online Journal of Distance Learning Administration, 5*(3). Retrieved from http://www.westga.edu/~distance/ojdla/fall53/valentine53.html.

VanDusen, G. C. (1997). The virtual campus: Technology and reform in higher education. *ASHE-ERIC Higher Education Report, 25*(5).

Wagner, E. (2005). Enabling mobile learning. *Educause Review, 40*(3). http://www.educause.edu/ir/library/pdf/erm0532.pdf.

Weller, M. (2007). Learning objects, learning design, and adoption through succession. *Journal of Computing in Higher Education, 19*(1), 26–47. https://doi.org/10.1007/BF03033418.

Wong, G. K. W. (2016). The behavioral intentions of Hong Kong primary teachers in adopting educational technology. *Educational Technology Research and Development, 64*(2), 313–338. https://doi.org/10.1007/s11423-016-9426-9.

Zanjani, N., Edwards, S. L., Nykvist, S., & Geva, S. (2016). LMS acceptance: The instructor role. *Asia-Pacific Education Researcher, 25*(4), 519–526. https://doi.org/10.1007/s40299-016-0277-2.

Learning Management Systems

3.1 Introduction

Teaching and learning have been carried out through the traditional face-to-face, time and place restricted mode, and it was considered to be the backbone of the educational system. Traditionally, teaching takes place in libraries or inside the classrooms on-campus by the teacher who is traditionally the center of knowledge. The technology-assisted teaching and learning model has been migrated from a traditional model to a distance model. The system of education is in the exploration phase of this new learning system and transforming from a traditional model to a d-learning, e-learning, and m-learning model. It is the future direction of e-learning. Learning management system (LMS) has become an essential software for faculty and students in their teaching and learning process. It is a computer software that facilitates electronic learning. The LMS is composed of computer software that incorporates functions for teaching, evaluating, and administrating courses. It is used for contents delivery, course registration, tracking, reporting, and also for administration. This chapter presents the importance of LMS, the historical perspective of LMS, types of LMS and its users, and key features of LMS.

3.2 Learning Management Systems (LMS)

The use of LMS is growing in higher education. Wide acceptance of a VLE (or LMS) highlighted the importance of the e-learning approach. LMSs such as WebCT and Blackboard are the components of digital e-learning and widely used for both on-campus and off-campus students at major universities of the world (Murshitha & Wickramarachchi, 2016; Nagy, 2016). Without a doubt, LMS is a cost-effective and convenient learning platform. Learning management system (LMS) is a computer software that facilitates F2F learning, e-learning, and handles all aspects of the learning process (Szabo & Flesher, 2002). It is also referred to as a virtual learning environment (VLE) (Shiau & Chau, 2015) or course management system (Chu et al., 2012). It incorporates functions for teaching, evaluating, administrating courses (Gilhooly, 2001) and makes them available to all students (Bogarín, Cerezo, & Romero, 2018; Rapuano & Zoino, 2006; Sharma et al., 2011). It is an excellent tool for students' evaluation and assessment abilities (Tortora, Sebillo, Vitiello, & D'Ambrosio, 2002). One of the most important functions of an LMS is an interaction between learners through chat, instant messaging, discussion board, email, and monitoring and tracking student activities (Rapuano & Zoino, 2006). It is an excellent tool for students' evaluation and assessment abilities (Tortora et al., 2002). According to Nichols (2003), an LMS is "a collection of e-learning tools available through a shared administrative interface". These include interaction tools (such as messaging, emails, discussion board, and virtual chat for communication with students) and course delivery (such as uploading a syllabus, course materials, assignments, and assessments). It helps instructors in designing student-centered courses that they can effectively deliver F2F in the classroom as well as virtual instructions online (Jordan & Duckett, 2018). Almost all institutions of higher education have adopted LMS to support e-learning (Alharbi & Drew, 2014; Nagy, 2016). For instance, in the United States, almost 100% of HEIs use LMS for teaching and learning (Lang & Pirani, 2014). LMS is now equally popular not only in the educational sector but also in governmental organizations and business sectors for the training of individuals in the business sector (Avgeriou, Papasalouros, Retalis, & Skordalakis, 2003). It helps corporate designers and educators to plan, organize, and deliver online courses in an effective way (Jafari, McGee, & Carmean, 2006). It has the potential of improving online education and creating a fully functional virtual classroom. Prior research indicates that the use of LMS is growing in higher education (Alghamdi, 2016; Gautreau, 2011; Nagy, 2016).

The LMS market is predicted to grow at a rate of 16% from 2016 to 2020. The adoption of LMS for web-based instruction continues to increase in higher education. It has

© Springer Nature Switzerland AG 2021
R. A. Khan and H. Qudrat-Ullah, *Adoption of LMS in Higher Educational Institutions of the Middle East*,
Advances in Science, Technology & Innovation, https://doi.org/10.1007/978-3-030-50112-9_3

generated new opportunities for educators and learners to interact at any time anywhere without physical boundaries. It can provide all the essential material and tools to manage virtual teaching and learning environments. The users of the LMS may be categorized into learners/students, instructors, and administrators (Avgeriou et al., 2003). The administrators are responsible for monitoring the operation of the LMS, to resolve the technical issues, and for administering the user accounts of the LMS. The instructors are responsible for creating their courses, presenting contents, evaluating their performance, interacting with the students, and providing feedback to the students. The learners are the key users of the system, the receivers of the content materials. The students can interact with each other and with instructors through asynchronous and synchronous communication tools (Avgeriou et al., 2003).

Antonenko, Derakhshan, and Mendez (2013) placed the existing industry practices of mobile learning into five levels:

Level 0: LMSs of this category are not prepared for mobile learning. *Level 1*: LMSs of this category (such as Moodle, Sakai) are graphically prepared for mobile devices. *Level 2*: This category (e.g. MLE-Moodle, MOMO, Blackboard MobileTM) includes mobile extensions. *Level 3*: LMSs of this category (e.g. BlackBerry PushcastTM) are stand-alone, self-sufficient mobile LMSs. *Level 4*: These are innovative mobile LMSs (e.g. cloud computing or touch-screen capabilities) and use new affordances of mobile devices.

3.3 A Historical Perspective on Learning Management Systems

Using LMS in teaching and learning is not a new concept; its history is connected back to 1960 and it was called an offline LMS because it was not supported by the web-based systems. Examples of LMSs include Blackboard, WebCT, eCollege, Moodle, Desire2Learn, Angel, and OPAL (Sharma et al., 2011). The description of some examples of LMSs is as follows:

- **PLATO™**: Programmed Logic for Automated Teaching Operations (PLATO) was the first computer-assisted learning system invented in 1960 by the staff of the University of Illinois at Urbana-Champaign (Plato website, 2014). Initially, the learners accessed PLATO through stand-alone computers, but later, PLATO made the courses online and available from anywhere.
- **Learning Manager™**: In 1980, a new LMS was introduced known as TLM™ (The Learning Manager™). TLM™ presented tools for developing, reporting, managing, and materials for learning. It became a popular

tool for e-learning management and could be accessed remotely by dial-up as a student or an instructor using a terminal emulator.

- **Andrew Project**: The Andrew Project was developed by Carnegie Mellon University in 1982. It offered a unified computing environment for online collaboration. The purpose of this system was to develop a platform for computer-aided teaching.
- **EKKO™**: The first version of the EKKO™ was a computer-based conferencing system, invented and implemented at NKI™. NKI Distance Education in Norway started online distance education courses in 1987. These courses were delivered through EKKO, NKI's self-developed LMS.
- **ATHENA**: This project was initially developed in 1983 and later evolved in 1990 at Massachusetts Institute of Technology University. ATHENA was a campus-wide networked system that was meant for writing, sharing, and communicating papers with other users. This instructional system included a tutor, a blackboard, a textbook, a simulator of complex systems, a virtual laboratory, a communication medium, and a special-purpose learning environment (Balkovich, Lerman, & Parmelee, 1985).
- **HyperCourseware™**: HyperCourseware™ was a prototype electronic educational environment invented at the University of Maryland by Kent Norman in 1990 to be used as an electronic classroom called the 'Teaching Theater'. The purpose of HyperCourseware™ was to give access electronically to teaching and learning resources such as lectures, textbooks, discussions, question and answer documents (Norman, 1994).
- **WebCT™**: It was presented in 1996 at the University of British Columbia by Murray Goldberg. The key features of WebCT™ were mail system, live chat, discussion boards, and downloadable materials and web pages (*WebCT*, 2011). WebCT™ was the first and most famous CMS used in higher education. In 2006 WebCT™ was acquired by Blackboard™ (*WebCT*, 2011).
- **Blackboard™**: Blackboard™ was originated in 1997 and offers software applications for learning and teaching. Blackboard™ is one of the most popular commercial LMS. It was founded in 1997. It has a head office in Washington, D.C. and other offices in North America, Europe, Asia, and Australia. It offered different platforms: Blackboard Learn™ (an LMS), Blackboard Connect™ (for time-sensitive information, e.g. text, email, voice, and social media), Blackboard Collaborate™ (for a virtual classroom for synchronous instruction), Blackboard Analytics™ (for leaders of institutions for self-service and easy access to important information), Blackboard Mobile™ (mobile version of the LMS), and Blackboard Transact™: it is meant for a secure off-campus and

on-campus shopping using their ID cards (Blackboard™, 2012).

- **Modular Object-Oriented Dynamic Learning Environment (Moodle)**: It is an open-source and a free CMS designed to help instructors in creating effective online teaching and learning environments (*Moodle*, 2010). Moodle is considered to be a high-value educational tool in higher education. It affords instructors the tools to promote and manage online learning. These tools include many features such as the delivery of lessons, forums, tests, modules, and surveys. As an open-source, Moodle has had excellent success after Blackboard™.
- **Desire2Learn™**: It was introduced in 1999. It provided an e-learning environment for HEIs, K-12 schools, businesses, and government. The Desire2Learn™ provided six different platforms: Learning Environment, Learning Repository, ePortfolio, Analytics, Mobile, and Capture.
- **Sakai™**: It was introduced by a special grant given in 2004 by the Mellon Foundation when the University of Michigan, Massachusetts Institute of Technology University, Indiana University, University of Berkeley, and Stanford University started constructing a common Courseware Management System.

3.4 Mobile Learning Management Systems

The growing acceptance of m-devices is one of the major reasons that LMS providers have included the compatibility of mobile devices in designing and developing their main product. Some companies made efforts in the design and development of LMS for a mobile platform. The firms that have released a mobile version of LMS include:

- **Moodle Mobile (MOMO)**: This open-source LMS includes Moodle Mobile (MOMO) features such as communications, forums, resources, mobile community, mobile offline learning objects (MLOs), and mobile blogging.
- **Blackboard Mobile Learn™** provides an interactive learning environment compatible with the mobile platform. Blackboard Mobile Learn™ enables teachers and students to get access to their content resources and courses on a variety of m-devices. Blackboard™ introduced mobile-learning applications on June 15, 2010 compatible with the majority of the mobile platforms such as Touch™, Android™, iPhone™, iPod, Blackberry™, and iPad™ (Blackboard™ website, 2010).
- **Desire2Learn 2GO™** was a mobile learning application, presented by Desire2Learn in July 2008 (McLeod, 2008). D2L 2GO™ was designed for Blackberry™ but was also compatible with other smartphones such as the

iPhone. These learning applications help both educators and learners to stay connected with each other anywhere and anytime. Later, in 2011, Desire2Learn™ launched Desire2Learn Campus Life™ that enabled users to collaborate, communicate, and share information (*Desire2Learn Campus Life*, 2011).

3.5 Open-Source and Commercial LMS

Many open-source and commercial LMSs are available in the market. The term 'open-source' means that individuals are allowed to use the program, study it, modify it, and redistribute copies free of charge (Nagy, 2016). The most popular open-source systems are MOODLE, CANVAS, Blackboard, and Electronic Educational Environment (EEE) LMS. Moodle is an open-source software that was developed as a Ph.D. research project in 1999 at the Curtin University of Technology in Western Australia by Martin Dougiamas. Other open-source LMSs include *ATutor*, developed by the University of Toronto; *ClassWeb*, developed by the University of California, Los Angeles; *Moodle*, developed by Martin Dougiamas, Australia; *Caroline*, developed by the Université Catholique de Louvain; *LON-CAPA*, developed by the Michigan State University, and *Coursework*, developed by the Stanford University. In the case of 'commercial LMS', a license should be purchased to use the product. Examples of commercial software include Blackboard Inc., Apex Learning and ANGEL Learning (Nagy, 2016).

3.6 Web-Based LMS Versus Installed LMS Software

One of the differences between 'web-based LMS' software and the 'installed LMS' software is that in installed LMS model, we have to install vendor-created LMS on one of the servers. It has to be maintained by some IT expert in our institution. It comes with a setup fee and maintenance arrangements. However, modern web-based online learning management systems are also available. One of the advantages of web-based LMS is that the pricing for web-based LMS is a fraction of the installed LMS. Also, the LMS vendors maintain the software and keep improving and developing their online LMS software. There are no huge setup fees needed. The monthly costs are also much lower than the installed LMS. There are two different views in the web-based LMS. It can be software as a service (SaaS) or you host the LMS software yourself. In the case of 'software as a service' (SaaS) model, a third party hosts the LMS applications and makes LMS available to institutions over the Internet. In the SaaS model, uptime and security-related issues are the responsibility of the LMS vendor. However, if you are hosting LMS software, you have full control of LMS software and the server. You are also responsible for the uptime and security-related issues of the LMS server.

3.7　Key Features of Learning Management Systems

LMSs offer ways to organize and manage learning and teaching resources for learners and educators. Although 'higher level' features (such as lesson activity module, Wiki activities, Blackboard collaboration, Blackboard Instructors, and Blackboard Apps for students) are available in many LMSs, the researcher compiled the most standard features used by most of the instructors (Jill, 2016; "LMS Software: Key Features," 2016; Woods, Baker, & Hopper, 2004).

The features used by most of the instructors include:

- **Instructional Features**: Instructional features of LMS include course creation, implementation, assignments, Gradebook, assessments, tests, and management of courses.
- **Content Management Features**: Instructors can manage and update all of their teaching material such as files, slides, PDFs, audio files, videos, images, and much more to a centralized location.
- **User Management Features**: User accounts and groups of users can be imported or exported into the LMS. User management includes creating and managing user accounts, importing and exporting user and group accounts, and maintaining user-profiles and password issues.
- **Interactive Features**: Interactive features of LMS consist of a discussion board, chat room, messaging, mutual uploading or downloading of files, digital drop boxes, and transfer of files between LMS and other application software (e.g. Microsoft Excel and Microsoft Word). Synchronous communication is one of the most important features of the LMS which is a real-time virtual classroom with an interactive whiteboard, application sharing, two-way communication, and file transferring. Discussion on LMS is another feature for posting questions and answers from the users on a discussion board. Similarly, instant messaging is sending and receiving text messages to other users and trainers of the LMS.
- **Visual Features**: Visual features of LMS deal with the visual appearance of the entire LMS platform. Visual features include graphic interfaces, colors, shapes of buttons, font types, font sizes, and linking of all elements with each other.

3.8　Summary

The use of LMS in teaching and learning has witnessed exponential growth because the Internet and smart devices have changed the millennial learners. This chapter also includes a description of various types of LMS and their functions used in education. A study and comparison of important features of LMS and the market of LMS are also carried out.

References

Alghamdi, S. R. (2016). Use and attitude towards Learning Management Systems (LMS) in Saudi Arabian universities. *Eurasia Journal of Mathematics, Science & Technology Education, 12*(9), 2309–2330. https://doi.org/10.12973/eurasia.2016.1281a.

Alharbi, S., & Drew, S. (2014). Using the technology acceptance model in understanding academics' behavioural intention to use learning management systems. *International Journal of Advanced Computer Science and Applications, 5,* 143–155.

Antonenko, P. D., Derakhshan, N., & Mendez, J. P. (2013). Pedagogy 2 go: Student and faculty perspectives on the features of mobile learning management systems. *International Journal of Mobile Learning and Organisation, 7*(3–4), 197–209. https://doi.org/10.1504/IJMLO.2013.057161.

Avgeriou, P., Papasalouros, A., Retalis, S., & Skordalakis, M. (2003). Towards a pattern language for learning management systems. *Educational Technology & Society, 6*(2), 11–24.

Balkovich, E., Lerman, S., & Parmelee, R. P. (1985). Computing in higher education: The Athena experience. *Communications of the ACM, 28*(11), 1214–1224.

Bogarín, A., Cerezo, R., & Romero, C. (2018). Discovering learning processes using Inductive Miner: A case study with Learning Management Systems (LMSs). *Psicothema, 30*(3), 322–329. https://doi.org/10.7334/psicothema2018.116.

Chu, L. F., Erlendson, M. J., Sun, J. S., Clemenson, A. M., Martin, P., & Eng, R. L. (2012). Information technology and its role in anaesthesia training and continuing medical education. *Best Practice & Research. Clinical Anaesthesiology, 26*(1), 33–53. https://doi.org/10.1016/j.bpa.2012.02.002.

Desire2Learn Campus Life. (2011). Desire2Learn announces release of latest mobile platform—Desire2Learn campus life. Retrieved from https://www.d2l.com/newsroom/releases/desire2learn-announces-release-of-latest-mobile-platform-desire2learn-campus-life/.

Gautreau, C. (2011). Motivational factors affecting the integration of a learning management system by faculty. *The Journal of Educators Online, 8*(1).

Gilhooly, K. (2001). Making e-learning effective: Industry trend or event. *Computerworld, 35*(29), 52–53.

Jafari, A., McGee, P., & Carmean, C. (2006). Managing courses defining learning: What faculty, students, and administrators want. *Educause Review, 4*(4), 50–51.

Jill, W. (2016). *The 10 Must-Have LMS Features.* https://www.skillbuilderlms.com/10-must-have-lms-features/.

Jordan, M. M., & Duckett, N. D. (2018). Universities confront 'Tech disruption': Perceptions of student engagement online using two learning management systems universities confront 'Tech disruption': Perceptions of student. *10*(1).

Lang, L., & Pirani, J. A. (2014). The learning management system evolution learning management systems: Now and beyond. In

Educause Annual Conference (pp. 1–9). Retrieved from http://www.educause.edu/annual-conference/2014.

LMS Software: Key Features. (2016). http://www.proprofs.com/c/lms/lms-systems-key-features/.

McLeod, J. (2008). Desire2Learn goes mobile with Desire2Learn 2GO for BlackBerry smartphones. Retrieved November 20, 2011 from http://www.desire2learn.com/news/newsdetails_87.asp.

Moodle. (2010). Moodle Docs. Retrieved from http://docs.moodle.org/en/.

Murshitha, S. M., & Wickramarachchi, A. P. R. (2016). A study of students' perspectives on the adoption of LMS at the University of Kelaniya. *Journal of Management, 9*(1), 16. https://doi.org/10.4038/jm.v9i1.7562.

Nagy, J. T. (2016). Using learning management systems in business and economics studies in Hungarian higher education. *Education and Information Technologies, 21*(4), 897–917. https://doi.org/10.1007/s10639-014-9360-6.

Nichols, M. (2003). A theory for eLearning. *Educational Technology & Society, 6*(2), 1–10.

Norman, K. L. (1994). Hypercourseware for interactive instruction in the electronic classroom. *Behavior Research Methods Instruments & Computers, 26*(2), 255–259. https://doi.org/10.3758/BF03204632.

Rapuano, S., & Zoino, F. (2006). A learning management system including laboratory experiments on measurement instrumentation.

Instrumentation and Measurement, IEEE Transactions On, 55(5), 1757–1766.

Sharma, S. A. T., Paul, A., Gillies, D., Conway, C., Nesbitt, S., Ripstein, I. R. A., … Mcconnell, K. (2011). Learning/Curriculum Management Systems (LCMS): Emergence of a new wave in medical education. *Learning, 11*(13).

Shiau, W. L., & Chau, P. Y. K. (2015). Understanding behavioral intention to use a cloud computing classroom: A multiple model comparison approach. *Information & Management, 53*(3), 355–365. https://doi.org/10.1016/j.im.2015.10.004.

Szabo, M., & Flesher, K. (2002). CMI theory and practice: Historical roots of learning management systems. *Educational Technology, 32,* 58–59.

Tortora, G., Sebillo, M., Vitiello, G., & D'Ambrosio, P. (2002). A multilevel Learning management system. In *14th International Conference on Software Engineering and Knowledge Engineering* (pp. 541–547). USA: ACM.

WebCT. (2011). http://en.wikipedia.org/w/index.php?title=WebCT&oldid=425085423.

Woods, R., Baker, J. D., & Hopper, D. (2004). Hybrid structures: Faculty use and perception of web-based courseware as a supplement to face-to-face instruction. *The Internet and Higher Education, 7*(4), 281–297.

4.1 Introduction

The adoption of new technology plays a vital role in shaping the future of the educational sector of any country (Mosa, Naz'ri bin Mahrin, & Ibrahim, 2016). Thus, it is important to understand the concept of users' adoption of technology. To improve the quality of education and to enhance the professional productivity of the instructors, the latest technologies are required to be adopted and integrated into the educational system (Tomei, 2005). The information technology used by the people is impacted by cultural values (Im, Hong, & Kang, 2011). Hence, many researchers argue that cultural variables need to be incorporated into technology acceptance models (Baptista & Oliveira, 2015; Lu & Lin, 2012). A sound understanding of Saudi cultural interaction with computing environments could lead to better acceptance of IT in the country (Alharbi, 2006). Thus, the study discusses the adoption of new technology in the cultural context of Saudi Arabia.

4.2 Study Settings: The Kingdom of Saudi Arabia

As our study was conducted in the Kingdom of Saudi Arabia (KSA), here we provide some background information about this country. The KSA was founded by King Abdul Aziz bin Abdulrahman Al-Saud in 1932. Being a Muslim state, it was assured by the government that the constitution would be governed by the Holy Quran and Sharia law. The most significant and attractive characteristics of the Kingdom of Saudi Arabia are the presence of the holy shrines in Makkah and Medina as well as it being the birthplace of Islam. These characteristics contribute to Muslim pilgrims visiting Saudi Arabia every year. Muslims all over the world turn their faces toward Holy Kaaba 'five times' at the time of every prayer. As shown in Fig. 4.1, the Kingdom of Saudi Arabia is the largest country in the region having an area of almost 2,150,000 square kilometers (868,730 square miles) that covers almost two-thirds of the Arabian Peninsula in the south-west of the continent of Asia ('The World Fact Book', 2016). There are seven Arab countries with its border: Jordan 731 km, Kuwait 221 km, Iraq 811 km, Oman 658 km, UAE 457 km, Yemen 1307 km, and Qatar 87 km ('The World Fact Book', 2016). Riyadh serves as the capital city and has a large population in the central region of KSA. According to the World Fact Book (2016), the average literacy rate of the total population is 94.4% (male: 97% and female: 91.1%). The currency of Saudi Arabia is Saudi Riyal. The economy depends mostly on the income from oil exports and petroleum products. Saudi ARAMCO is the state-owned petroleum company. The natural resources of KSA are petroleum, natural gas, iron ore, gold, and copper (The World Fact Book, 2016).

Saudi Arabia is a developing country where technology acceptance on an individual and business level is highly influenced by government policies (Alharbi, 2006). When the Internet was launched into KSA, a prime concern for the government was about undesirable material and contents, such as pornography. Therefore, an internet filter managed by the King Abdulaziz City of Science and Technology was established in the city of Riyadh to control undesirable contents. The Kingdom of Saudi Arabia (KSA) is focusing on the adoption of new technology (such as cloud computing) due to its vital benefits and their extensive use in various organizations (Scherer, Siddiq, & Tondeur, 2019). The adoption of new technology plays a significant role in shaping the future of the educational sector of any country (Mosa et al., 2016). Thus, it is important to understand the concept of users' adoption of technology.

4.3 Technology Adoption in Saudi Context

Information and communication technology (IC&T) is rapidly growing in Saudi Arabia. The Saudi government has taken various steps to transform Saudi Arabia into a digital society (Aljabre, 2012). The number of Internet users grew

© Springer Nature Switzerland AG 2021
R. A. Khan and H. Qudrat-Ullah, *Adoption of LMS in Higher Educational Institutions of the Middle East*,
Advances in Science, Technology & Innovation, https://doi.org/10.1007/978-3-030-50112-9_4

Fig. 4.1 Kingdom of Saudi Arabia map. *Source* http://www.mapsofworld.com/saudi-arabia/

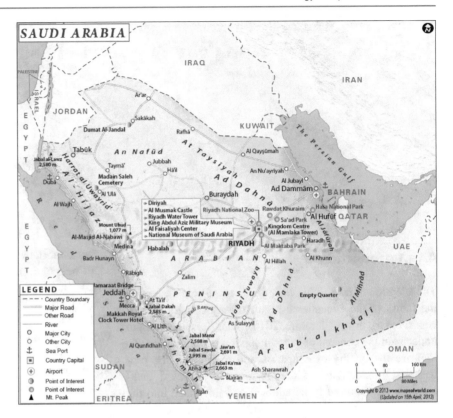

from around 1 million in 2001 to 20 million by the end of 2016 (*Internet World Stats*, 2017). This rapid development in technology is credited to the reduced prices of computers and the Internet. Also, the Internet brought Arabic language sites, and e-services available in most of the official websites. The Kingdom of Saudi Arabia has made a huge investment to improve the nation's educational system (Aljabre, 2012). Despite the recent expansion and development of higher educational institutions, there is still insufficient room in the HEIs to accommodate a large number of students (Al-Shehri, 2010). Every academic year, thousands of applicants are left without a place at their desired institutions to study their subject of preference. Another fact is that KSA is a large country with problems of accessibility, particularly for women and those who are unable to travel to the places where the institutions are located (Al-Shehri, 2010). A part of the solution to this issue is contained in d-learning (Al-Shehri, 2010).

Distance learning has been present in higher educational institutions (HEI) in many countries like the UK and the USA for some time (Aljabre, 2012). The Arab Open University (AOU) in most of the Arab countries (http://www.arabou.org.sa) is a popular institution that follows a hybrid e-learning model. It involves having a physical campus where learners go for meeting their educators and for administrative work. Currently, the headquarter of AOU is in Kuwait, and other branches of the AOU are in KSA, Lebanon, Bahrain, Egypt, Jordan, Sudan, and Oman.

Hamdan Bin Mohammed e-University (http://www.hbmeu.ac.ae) was assigned a project by the Ministry of Higher Education to set criteria for the adoption of e-learning for other universities in the region (Al-hunaiyyan, Al-hajri, Alzayed, & Alraqqas, 2016). The Saudi government recognized the need for integrating ICT at HEIs (Alenezi, 2012). The use of distance education is also growing in KSA that opened a new model of education (Aljabre, 2012). Ministry of HE (MoHE) is among those organizations that suggested the use of online learning and d-learning systems in Saudi Arabia (Alenezi, 2012). The following governmental organizations were involved with this project in KSA: Ministry of Telecommunication, Ministry of Higher Education, National Center of E-learning, King Abdulaziz City for Science and Technology (KACST), and some HEIs (Al-shehri, 2010). In this regard, three universities: King Abdulaziz University in (KAU) Jeddah, King Saud University (KSU) in Riyadh, and King Faisal University (KFU) in Al-Hasa were the leading universities to start work on a d-learning project in KSA. The mission of the universities' distance-learning project was to integrate, utilize, and train instructors and students with the state-of-the-art technology in d-learning. The d-learning and e-learning projects at each of these institutions were established at different times with KAU being the first in 2006. In 2010, both KFU and KSU initiated a new Deanship of e-learning and d-learning at their universities. The collective vision of these three universities was to give a recommendation related to the risks and opportunities related to e-learning in the future of Saudi HEIs. The systems that are made available to the learners included Web CT and Blackboard as LMSs. The Blackboard system can fully support a distance learning environment or act as a supplemental site for face-to-face (F2F) learning or can provide a hybrid course through a blended learning environment. The advantages of online programs are that they can be used by children, the young, the old, parents, working professionals, and the handicapped. The use of d-learning has produced new opportunities for education for Saudis, especially women.

However, there are some challenges associated with d-learning and e-learning in Arab countries. Traditional classroom teaching and learning has been the norm for years and to ask students not to come to class may feel strange and perhaps difficult to accept. Online learning in Arab countries is still immature compared to developed countries (Tarhini, Hone, & Liu, 2014) since institutions still support traditional styles of F2F teaching due to the lack of trained staff, telecommunication infrastructure, government policies, and financial issues (Al-hunaiyyan et al., 2016). The 'absence' of the teacher, who is often considered as a guru and a single source of knowledge, might be a cause of concern for students who may fear to deal with certain aspects of distance learning. Therefore, the implementation of online learning at HEIs has been approached with some trepidation and resistance from both instructors and students (Aljabre, 2012) due to its lower status than F2F teaching. Hence, a majority of students in HEIs are still reluctant to use online courses (Simeonova, Bogolyubov, & Blagov, 2010). The majority of the students are unwilling to enroll in and often withdraw from online courses for various reasons. A large number of dropout students from online courses are choosing traditional F2F classroom settings (Ibrahim, Rwegasira, & Taher, 2007). Another reason might be the fact that compared to developed countries, the Internet is not easily accessible by all students as the infrastructure for the Internet is still not well-built in all regions of Saudi Arabia. One of the most important barriers to the willingness of various Arab universities to adopt online learning is that web-based degrees are not always accepted, and candidates will consequently have fewer job opportunities compared to traditional F2F degrees (Al-hunaiyyan et al., 2016). Although the use of e-learning and d-learning is gradually growing in Saudi Arabia, there are still many technical, social, and cultural challenges in technology adoption that the Saudi HEIs are facing.

4.4 Culture and Technology Adoption

The cultural and social characteristics of Arab societies and Muslim communities are different compared to Western culture. Especially, Saudi Arabia is a kind of traditional country where Islamic values and Arabian cultures are dominant (Alharbi, 2006). All traditions, cultural values, patterns, community practices, and social standards are determined by Islam (Al-saggaf, 2004). According to Al-Saggaf (2004), the culture is strongly influenced by religion in Saudi Arabia and it influences in all aspects of life. The Kingdom of Saudi Arabia is an epitome of the Middle Eastern culture. Thus, it is important to understand the influence of culture on technology adoption (Baptista & Oliveira, 2015). A sound understanding of Saudi cultural interaction with computing environments could lead to better acceptance of IT in the country (Alharbi, 2006).

Culture is a complex and broad concept, which can be defined in many different ways (Baptista & Oliveira, 2015). Gallivan and Srite (2005) concluded that culture is not easily observable but rather connected with values, meanings, and norms; therefore, it is difficult to access it. In the literature, there are numerous definitions of culture related to beliefs, ideologies, traditions, practices, norms, and other elements. O'Sullivan, Hartley, Saunders, Montgomery, and Fiske (1994) describe culture as the development of social sense and awareness. According to Hofstede (1980b), the word culture includes all habits, religions, traditions, arts, and languages and acts as a system of collective values that differentiates the individual of one group from another. It involves at least three components: what people think, what they do, and the material and products they produce (Boldley, 2004). Culture is also defined as "the integrated pattern of behavior that includes thought, speech, action, and artifacts, and it depends on man's capacity for learning and transmitting knowledge to succeeding generations" (Adler, 2001, p. 23).

In general, there are two approaches to examining cultural values: *emic* and *etic* (Morris, Ames, & Lickel, 1999). The emic approach includes the process that examines the culture from inside. In this approach, the researchers wish to examine the behavior of individuals in a society from their point of view. The etic approach "describes the ways in which how cultural variables fit into general causal models of a particular behavior" (Morris et al., 1999, p. 783). In this approach, the researchers use predetermined categories to investigate selected aspects of the cultures under examination. Van Oudenhoven (2001) states that there is a difference between organizational cultures and national culture. The national culture is linked with the values, beliefs, and practices shared by the group of individuals related to a nation, while the organizational culture is concerned with beliefs, values, and practices shared by most individuals of an organization. The individuals together make a culture (Zakour, 2004) which affects adoption and usage behaviors, and it must be considered when studying the adoption of any technology.

4.5 Culture Theories

Many approaches have been considered to investigate the effects of cultural dimensions including Hofstede's (1980a) cultural dimensions, social identity theory (Straub, 2002), and Triandis's (1982) cultural dimensions. Hofstede's (1980a) cultural dimensions are the most widely recognized and accepted for comparison of cultures on *national levels* in multiple countries. Many researchers adopted Hofstede's questionnaire to study culture at the *individual level*. Baptista and Oliveira (2015) used Hofstede's (1980a) cultural dimensions to understand the individual and situational characteristics in the adoption of mobile banking, adding new understandings into how culture impacts an individual's behavioral intention. Further, Hofstede's work has also been well appreciated and accepted in investigating the influence of culture on technology acceptance (Straub, Keil, & Brenner, 1997). Several studies are available in which researchers have developed different types of cultural classification systems. There are similarities in some cases and there is also overlap in a few cases. Three well-known cultural classification schemes are: (a) Schwartz's value inventory, (b) Inglehart's world values survey, and (c) Hofstede's value survey module.

- **Schwartz's Value Inventory**: A different approach was used to determine the values of a culture. Hofstede's survey asked for a preferred outcome, but Schwartz asked to rank the importance of certain values from 60,000 participants, not just from business areas, but from all areas of endeavor and outlined ten cultural values in two groups (Schwartz, 1992). Group 1 was named 'Individual Interest' and included power, hedonism, achievement, self-direction, and stimulation. Group 2 was named 'Collective Interest' and included universalism, tradition, benevolence, security, and conformity. The results from Schwartz's study were not widely dispersed and did not have as appropriate a population sample as Hofstede (1980a).
- **Inglehart's World Values Survey**: This study collected and analyzed the data from 80 countries for two indices.

The first category was named 'Traditional-to-Secular/Rational values' and shows the difference in countries from traditional family ties, religion, and deference to authority points of view. The second category was named 'Survival-to-Self-Expression values' and reflects the desire of countries for basic survival (i.e. shelter and food) against the desire for equal rights and self-expression.

- **Hofstede's Value Survey Module**: Hofstede's value survey is one of the most prominent cultural classifications that involved more than 100,000 IBM employees from more than 50 countries between 1967 and 1973. The initial IBM questionnaire was developed into the form called VSM 80 (Values Survey Module, developed in 1980) that was further updated into VSM 81 and VSM 82. The initial versions of VSM included questions related to business populations. For instance, the word 'boss' was used in early versions; this word was invalid for the general population and later it was corrected in VSM 94. A later version of VSM includes VSM2008 and VSM 2013.

4.6 Hofstede's Cultural Dimensions and Saudi Arabia

Hofstede (1980a) identified four indices by which a culture can be measured (later more dimensions were included). The description of all dimensions is described as follows:

- **Power Distance Index (PDI)**: The PDI refers to inequality among the people of the country's culture. The PDI explains the degree to which the individuals of a society accept less power (Hofstede, 2011). Greater PDI means that inequalities of power and wealth are evident in a culture (Al-Gahtani et al., 2007). The cultures with a greater PDI index have authority differentials and unequal distribution of power, and centralized decision shapes their main societal culture (Hofstede, 1980a). High PDI cultures exhibit a hierarchical structure where the individuals are very much concerned about agreeing with their bosses' opinions. In cultures having high power distance, some individuals will be seen as more influential or important. The employees in an Arab culture demonstrate a significant relationship between the variables of social influence and behavioral intention, compared, for instance, with the US (Al-Gahtani et al., 2007).
- **Individualism (INDV) versus Collectivism**: The Individualism (INDV) describes the amount of individualism of a society. In individualistic cultures, the relations between people are not strong enough and the individuals of the society are supposed to take care of themselves;

however, in collectivistic societies, individuals are united into cohesive groups. With less INDV value, societies display collective decisions leading to a slow adoption decision process. The studies revealed that in the cultures with greater INDV, it is easier to adopt new technologies (Van Everdingen & Waarts, 2003). According to Hofstede (2010), in individualistic societies, the relationships among individuals are not strong and the people of that culture tend to take care of themselves, while in collectivistic cultures, people are unified into united groups. Lesser values of the individualism index (IND) indicate closer ties among people and the culture is recognized as having collective decisions that lead to a slow process of technology adoption.

- **Uncertainty Avoidance**: Uncertainty avoidance (UA) refers to the level of tolerance for ambiguity and uncertainty within the culture. The UAI dimension shows to what degree a society programs its individuals to feel either comfortable or uncomfortable in unstructured circumstances. A high UAI reflects a rule-oriented and structured society having rules, regulations, and controls. In greater UAI values, societies resist accepting any innovations. A culture with a high UAI is not able to take risks to accept innovations and will adopt innovations that have been already used by others (Im et al., 2011). Consequently, it will not be easy to adopt innovations in a culture with a greater UAI index (Van Everdingen, & Waarts, 2003).

- **Masculinity (MASC) versus Femininity:** The MAS dimension shows the distribution role of gender within a culture (Hofstede, 1980a). In a high masculinity culture, people are more goal-oriented. The MAS society is known for recognition for improvement, competition, more task-oriented, achievement, and success of the people; and these characteristics of culture lead to the adoption of new technology (Van Everdingen, & Waarts, 2003). On the other hand, in feminine culture, people are more people-oriented and therefore they pay more attention to others' opinions (Sun & Zhang, 2006). A low score on this dimension shows that individuals in the society are caring for others and there is a quality of life. In other words, managers in less MAS index culture to seek consensus on majority tasks (Al-Gahtani et al., 2007).

- **Long (LTO) and Short (STO) Time Orientation**: A fifth index (Bond, 2010) was added later as long- versus short-term orientation (LTO). This dimension was included by Michael Bond, a psychologist at the Chinese University of Hong Kong. Bond determined that some of the dimensions are not related to Chinese culture and recommended the inclusion of '*Confucian Dynamism*' which was recognized as long-term and was integrated into VSM 94 as LTO (long-term orientation). Bond's

original research was made for students from 23 different countries focused on comparing Chinese culture with other cultures. The LTO refers to the society that holds long-term aspirations and forward-thinking values. LTO is related to the culture's time prospect which signifies the importance linked to the future, the past, and the present. Cultures characterized by an LTO are focused on future rewards whereas an STO is typified by a focus on the short-term rewards of the past and the present (Al-Gahtani et al., 2007).

- **Indulgence**: Recently, the two new dimensions that have been included are indulgence vs restraint index (IVR) and monumentalism (MON) index (Hofstede, 2010). Both of these dimensions are experimental at the time of this research, and for this reason, will not be taken into account.

4.6.1 Criticism of Hofstede's Dimensions

There has been a broad criticism of Hofstede's (1980a) study of national culture. Some researchers have criticized the Hofstede's survey instrument because it contains culturally sensitive and subjective questions (Schwartz, 1999). Another major criticism of the study was the 'one company approach'. Most researchers considered one company's employees (i.e. IBM) may not be the best representatives of the national culture (Sondergaard, 1994). In other words, 'organizational culture may be so strong that national culture traits are overshadowed. Ignoring the potential interactions between the two may lead to erroneous conclusions' (Nakata & Sivakumar, 1996). According to Hofstede (1980a), cultural values and beliefs can be measured with the national identity. Hofstede's study was criticized by many researchers that measuring only national culture may not be effective 'ways of' analysis because cultures are not always confined by geographic edges but are generally represented within multiple nations (DiMaggio, 1997). Some researchers have criticized that the population in Hofstede's study was homogenous and there was little of the variety found in true populations. It was also criticized for overseeing the variations in communities (Smith & Bond, 1998). Hofstede (1980a) grouped all Arab countries into one cluster as a homogeneous culture mainly due to the commonalities of language and religion. This approach became more problematic when Hofstede ignored all cultural differences and assumed that all Arabs have the same ranking and scoring. He grouped many countries into clusters based on the commonalities between the countries. He grouped Latin, Arabs, Asian and Anglo into one cluster mainly due to language and religion. Williamson (2002) recommended observing the individual's values, behaviors, and differences that make a culture. However, Hofstede calculated and provided different values for every country within a cluster. For instance, the Anglo cluster included Canada, USA, Australia, UK, South Africa, Ireland, and New Zealand. The power distance of the UAS is 91 whereas other countries have different scores indicating cultural variation despite the commonality of religion and language. Similarly, all Arab countries were allotted similar scores by Hofstede for all dimensions in the earlier Hofstede (1980a) model. Clustering all Arab regions together led to incorrect measurements by overlooking at the small level cultural differences between Arab countries. Having commonalities that share the same religion and language does not justify clustering into one group (Williamson, 2002). Al-Khatib, Vitell, Rexeisen, and Rawwas (2005) also challenged the supposition of gathering Arabs in one cluster to make a homogenous group. Kalliny and Gentry (2007) warned that basing similarities on Islam, as there are many, does not mean that Arab Muslims and the Americans Muslims have a similar culture. It was suggested that research should be carried out for individual countries to get deeper insights rather than relying on the clustering of countries. Western researchers such as Hofstede claimed that all 22 Arab countries are homogenous in their culture, while other studies show that there are differences in their values, attitudes and other cultural elements, but the researchers have failed to provide depths and insights about the adoption of technology.

4.6.2 Strengths of Hofstede's Dimensions

Despite all criticism, Hofstede received huge support from many researchers. When he published in 1980, not many other studies were published on culture. Hofstede's study is considered a pioneer for many (Sondergaard, 1994). Many

researchers give credit to Hofstede for opening people's eyes to the significance of culture (Baptista & Oliveira, 2015). According to Sondergaard (1994) in the majority of cases, the results of Hofstede's studies were found to be matched with other researchers. Hofstede's dimensions are the most widely recognized and accepted for comparison of cultures on a national-level and multiple countries. Hofstede's research on cultural dimensions has been extensively used and cited in many studies by many other researchers.

4.7 The Comparison of Hofstede's Dimensions Between Saudi Arabia and the UK

Figure 4.2 shows a comparison of Hofstede's dimension of KSA with the scores of the United Kingdom (UK). A comparison of scores of KSA with other countries would help in understanding the meaning of low or high values of the cultural dimensions of KSA. The UK has been selected in comparison with KSA because of the high difference in values of most cultural dimensions.

As shown in Fig. 4.2, PD of Saudi Arabia is high (95) as compared to the UK (35). This implies that in a high PD society the subordinates are told (ordered) to do their jobs and the boss is an autocrat (Al-Gahtani et al., 2007). The IND value for Saudi Arabia is 25, whereas the IND value of the UK is 89. The low value of the individualism index for KSA refers to a culture where collective opinions are more important. The score of UA for KSA is 80, whereas it is 35 for the UK. The countries demonstrating such a high value of UA retain rigid codes of behavior and belief. The innovations and adoptions of technology are resisted in such societies. The score for KSA is 60 on the MAS dimension showing that KSA is a masculine society. The individuals in

Fig. 4.2 The comparison of Hofstede's dimensions between Saudi Arabia and the UK. *Source* https://geert-hofstede.com/saudi-arabia.html

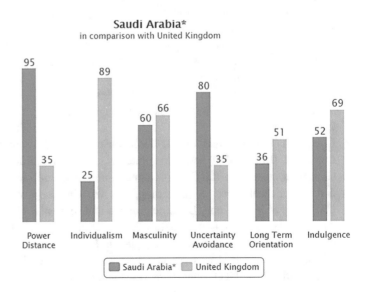

such cultures are more concerned with sticking with the absolute truth and are considered as normative in their thinking. Such individuals demonstrate more admiration for traditions and are focused on achieving results in a quicker time. The LTO value of KSA is 36 (whereas the UK is 51). The cultures having a low value of LTO (36) show low technology adoption (Al-Gahtani et al., 2007). As shown in Fig. 4.2 that KSA has scores of 52 and the UK has 69 on the indulgence dimension. This score for KSA is in the middle and does not indicate a clear distinction of this dimension. Hofstede (2010) discovered that in Saudi Arabia the PD value (95) and UA value (80) are very high. These values reflect that the Saudi society as a whole is not ready to accept any kind of change and risk. Furthermore, Saudi Arabia is extremely rule-oriented having strict laws and rules to minimize the degree of uncertainty experienced.

4.8 Summary

Adoption of technology is a cultural issue. The organizations using information systems are affected by a culture that should be taken into consideration before integrating any new technology in any organization (Im et al., 2011). Many researchers argue that cultural variables need to be incorporated in technology acceptance models (Baptista & Oliveira, 2015; Lu & Lin, 2012) because information technology used by the people is impacted by cultural values (Im et al., 2011), and the strength of any model or theory can be established by considering it across diverse cultures (Al-Qeisi, 2009). In this regard, Hofstede (1997) argued that culture has a remarkable influence on user-behavior and cultural dimensions deliver a theoretical foundation for discovering the adoption of technology. Therefore, in line with other researchers, such as (AL-Somali & Abdullah, 2012; Nistor, Lerche, Weinberger, Ceobanu, & Heymann, 2014), this study extends the UTUAT2 model with the Hofstede's cultural dimensions. Thus, this chapter examined the adoption of technology in the cultural context of higher educational institutions. We also discussed Saudi culture, cultural theories, Hofstede's cultural dimensions, and technology adoption models.

References

Adler, N. (2001). *International dimensions of organizational behavior.* South-Western Publication.

Al-Gahtani, S. S., Hubona, G. S., & Wang, J. (2007). Information technology (IT) in Saudi Arabia: Culture and the acceptance and use of IT. *Information and Management, 44*(8), 681–691. https://doi.org/10.1016/j.im.2007.09.002.

Al-hunaiyyan, A., Al-hajri, R., Alzayed, A., & Alraqqas, B. (2016). Towards an effective distance learning model: Implementation framework for Arab universities. *International Journal of Computer Application, 6*(5), 87–102.

Al-Khatib, J. A., Vitell, S. J., Rexeisen, R., & Rawwas, M. (2005). Inter-country differences of consumer ethics in Arab countries. *International Business Review, 14*(4), 495–516. https://doi.org/10.1016/j.ibusrev.2005.04.001.

Al-Qeisi, K. (2009). Analyzing the use of UTAUT model in explaining an online behavior: Internet banking adoption [Brunel University]. http://bura.brunel.ac.uk/bitstream/2438/3620/1/KholoudThesis.pdf.

Al-saggaf, Y. (2004). The effect of online community on offline community in Saudi Arabia. *Ejisdc,* 1–16.

Al-Shehri, A. M. (2010). E-learning in Saudi Arabia: "To E or not to E, that is the question". *Journal of Family & Community Medicine, 17*(3), 147–150. https://doi.org/10.4103/1319-1683.74333.

AL-Somali, & Abdullah, S. (2011). *Electronic commerce adoption: A study of business-to-business practices in saudi arabia.* Aston University.

Alenezi, A. R. (2012). E-learning acceptance : Technological key factors for successful students' engagement in E-learning system. In *EEE'12-The 2012 International Conference on e-Learning* (pp. 16–19).

Alharbi, S. J. M. (2006). *Perceptions of faculty and students toward the obstacles of implementing E-Government in educational institutions in Saudi Arabia.* West Virginia: Morgantown.

Aljabre, A. (2012). An exploration of distance learning in Saudi Arabian universities : Current practices and future possibilities, *2*(2), 132–138.

Baptista, G., & Oliveira, T. (2015). Understanding mobile banking: The unified theory of acceptance and use of technology combined with cultural moderators. *Computers in Human Behavior, 50,* 418–430. https://doi.org/10.1016/j.chb.2015.04.024.

Boldley, J. H. (2004). *Cultural anthropology: Tribes, states, and the global system, with powerweb.* McGraw-Hill.

Bond. (2010). *Michael Harris bond.* Retrieved from http://michael.bond.socialpsychology.org.

DiMaggio, P. (1997). Culture and cognition. *Annual Review of Sociology, 23*(1), 263–287.

Gallivan, M., & Srite, M. (2005). Information technology and culture: Identifying fragmentary and holistic perspectives of culture. *Information and Organization, 15*(2), 295–338. https://doi.org/10.1016/j.infoandorg.2005.02.005.

Hofstede, G. (1997). *Cultures and organizations: Software of the mind.* McGraw Hill.

Hofstede, G. (1980a). *Culture's consequences: International differences in work- related values.* Sage Publications.

Hofstede, G. (1980b). Culture and organizations. *International Studies of Management & Organization, 10*(4), 15–41.

Hofstede, G. J. (2010). *Cultures and organizations- software of the mind: Intercultural cooperation and its importance for survival (Revised an).* McGraw Hill.

Hofstede, G. (2011). Dimensionalizing cultures : The Hofstede model in context. *Online Readings in Psychology and Culture, 2*(1), 1–26. https://doi.org/10.9707/2307-0919.1014.

Ibrahim, M., Rwegasira, K., & Taher, A. (2007). Institutional factors affecting students' intentions to withdraw from distance learning programs in the Kingdom of Saudi Arabia: The case of the Arab Open University. *Online Journal of Distance Learning Administration, 10*(1).

Im, I., Hong, S., & Kang, M. S. (2011). An international comparison of technology adoption. *Information & Management, 48*(1), 1–8. https://doi.org/10.1016/j.im.2010.09.001.

Internet World Stats. (2017). *Internet usage, broadband and telecommunications reports.* Retrieved from http://www.internetworldstats.com/me/sa.htm.

Kalliny, M., & Gentry, L. (2007). Cultural values reflected in Arab and American television advertising. *Journal of Current Issues & Research in Advertising, 29*(1), 15–32.

Lu, H. -K., & Lin, P. -C. (2012). Toward an extended behavioral intention model for e-learning: Using learning and teaching styles as individual differences. In *2012 2nd International Conference on Consumer Electronics, Communications and Networks (CECNet)* (pp. 3673–3676). https://doi.org/10.1109/CECNet.2012.6202261.

Morris, M. W., Ames, D., & Lickel, B. (1999). Views from the inside and outside: Integrating Emic and Etic insights about culture and justice judgment. *Academy of Management Review, 24*(4), 781–796. https://doi.org/10.5465/AMR.1999.2553253.

Mosa, A. A., Naz'ri bin Mahrin, M., & Ibrrahim, R. (2016). Technological aspects of e-learning readiness in higher education: A review of the literature. *Computer and Information Science, 9*(1), 113. https://doi.org/10.5539/cis.v9n1p113.

Nakata, C., & Sivakumar, C. K. (1996). National culture and new product development: An integrative review. *Journal of Marketing, 60*(1), 61–72.

Nistor, N., Lerche, T., Weinberger, A., Ceobanu, C., & Heymann, O. (2014). Towards the integration of culture into the unified theory of acceptance and use of technology. *British Journal of Educational Technology, 45*(1), 36–55. https://doi.org/10.1111/j.1467-8535.2012.01383.x.

O'Sullivan, Hartley, J., Saunders, D., Montgomery, M., & Fiske, J. (1994). *Key concepts in communication and cultural studies.* London and New York: Routledge.

Scherer, R., Siddiq, F., & Tondeur, J. (2019). The technology acceptance model (TAM): A meta-analytic structural equation modeling approach to explaining teachers' adoption of digital technology in education. *Computers & Education, 128*(0317), 13–35. https://doi.org/10.1016/j.compedu.2018.09.009.

Schwartz, S. (1999). A theory of cultural values and some implications for work. *Applied Psychology, 48*(1), 23. https://doi.org/10.1080/026999499377655.

Schwartz, S. H. (1992). Universals in the content and structure of values: Theoretical advances and empirical tests in 20 countries. In M. P. Zanna (Ed.), *Advances in experimental psychology* (Vol. 25). Academic Press.

Simeonova, B., Bogolyubov, P., & Blagov, E. (2010). Use and acceptance of learning platforms within universities. *The Electronic Journal of Knowledge Management, 12*(1), 26–37. Retrieved from www.ejkm.com.

Smith, P. B., & Bond, M. H. (1998). *Social psychology across cultures.* Prentice Hall.

Sondergaard, M. (1994). Research note: Hofstede's consequences: A Study of reviews citation and replications. *Organization Studies, 15*(3), 447–456.

Straub, D., Keil, M., & Brenner, W. (1997). Testing the technology acceptance model across cultures: A three country study. *Information & Management, 33*(1), 1–11.

Straub, D. (2002). Toward a theory-based measurement of culture. *Journal of Global Information Management, 10*(1), 24–32.

Sun, H., & Zhang, P. (2006). The role of moderating factors in user technology acceptance. *International Journal of Human Computer Studies, 64*(2), 53–78. https://doi.org/10.1016/j.ijhcs.2005.04.013.

Tarhini, A., Hone, K., & Liu, X. (2014). The effects of individual differences on e-learning users' behaviour in developing countries: A structural equation model. *Computers in Human Behavior, 41* (153–163).

The World Fact Book. (n.d.). The Central Intelligence Agency (CIA) Library, Saudi Arabia,. Retrieved July 31, 2016, from https://www.cia.gov/library/publications/the-world-factbook/geos/sa.html.

Tomei, L. A. (2005). *Taxonomy for the technology domain.* Information Science Publishing.

Triandis, H. C. (1982). Dimensions of cultural variation as parameters for organizational theories. *International Studies of Management and Organization, 12*(4), 139–159.

Van Everdingen, Y. M., & Waarts, E. (2003). The effect of national culture on the adoption of innovations. *Marketing Letters, 14*(3), 217–232. https://doi.org/10.1023/A:1027452919403.

Van Oudenhoven, J. P. (2001). Do organizations reflect national cultures? A 10-nation study. *International Journal of Intercultural Relations, 25*(1), 89–107.

Williamson, D. (2002). Forward from a critique of Hofstede's model of national culture. *Human Relations, 55*(200211), 1373–1395. https://doi.org/10.1177/00187267025511006.

Zakour, A. (2004). Cultural differences and information technology acceptance. *SAIS 2004 Proceedings.*

5.1 Introduction

This research compares different technology adoption models, and then builds a model on the modified unified theory of acceptance and use of technology (UTAUT2). This study extends the UTAUT2 model with 'technology awareness' and 'Hofstede's cultural dimensions' as moderating variables of the UTAUT2 model. This study will also examine the adequacy of the original UTAUT2 model in the higher educational institutions (HEI) of Saudi Arabia.

Numerous theories and models have been presented to examine the variables impacting the adoption of new technologies (Baptista & Oliveira, 2015; Van Biljon & Kotzé, 2008; Rodrigues, Sarabdeen & Balasubramanian, 2016). According to Oliveira and Martins (2011), many technology acceptance theories exist in information systems research. Some popular technology adoption theories and models consist of: theory of planned behavior, motivational model, decomposed theory of planned behavior, theory of reasoned action, technology acceptance model, combined TAM, and TPB, diffusion of innovation, TAM2, TAM3, model of PC utilization, social cognitive theory, unified theory of acceptance and use of technology (UTAUT), and extended UTAUT (Alazzam, Basari, Sibghatullah, Ibrahim, Ramli, & Naim, 2016; Brown & Venkatesh, 2005; Ghobakhloo, Zulkifli, & Aziz, 2010; Jayasingh & Eze, 2010). The models such as technology, organization, and environment (TOE) and diffusion of innovation (DOI) deal with the adoption of technology at the *firm level*. The technology acceptance theories that deal with acceptance at the *individual level* include technology acceptance model (TAM), the theory of planned behavior (TPB), and the unified theory of acceptance and use of technology (Rahim, Lallmahomed, Ibrahim, & Rahman, 2011). This research will explore and adopt the model that deals with technology acceptance at the individual level.

5.2 Behavioral Intention (BI) and Use Behavior (UB)

This study reviews various models of technology adoption and the constructs of the models. However, it is important to understand the intentions and behaviors of the users for the adoption of technology. A fundamental concept behind adoption is that the '*intention*' of a person to adopt new technology is the prediction of its '*actual usage*' (Ajzen, 1991; Davis, Bagozzi & Warshaw, 1989; Venkatesh & Davis, 2000). Many researchers such as Compeau and Higgins (1995), Taylor and Todd (1995), Davis et al. (1989), and Venkatesh and Davis (2000) have been using 'intention to use' and 'actual usage' as dependent variables of the technology adoption. The use of behavioral intention (BI) and actual use behavior (UB) have been used interchangeably by the researchers as dependent variables. Previous studies show that the 'intention to use' is a predictor of 'use behavior' of a technology (Ajzen, 1991; Sheppard, Hartwick, & Warshaw, 1988). Venkatesh, Morris, Davis and Davis (2003) also concluded that user's acceptance of new technology is dependent on 'the behavioral intention' and 'actual use' of technology. Literature shows that these two determinants are considered to measure the degree of acceptance of a technology (Keramati, Sharif, Azad, & Soofifard, 2012). A review of the literature shows that the 'intention to use' a technology is highly correlated with the 'actual use' of the technology (Shiau & Chau, 2015; Wakefield & Whitten, 2006). Figure 5.1 demonstrates the most basic model of 'information system acceptance', adapted from Venkatesh et al. (2003). It shows that an individual's reactions to use technology make the individual's intention directly linked to its actual use. The dotted line in the following figure is the feedback loop from usage and reflects a user's continued intention of using the system. Hence, the 'behavioral intention' (BI) has a significant

© Springer Nature Switzerland AG 2021
R. A. Khan and H. Qudrat-Ullah, *Adoption of LMS in Higher Educational Institutions of the Middle East*,
Advances in Science, Technology & Innovation, https://doi.org/10.1007/978-3-030-50112-9_5

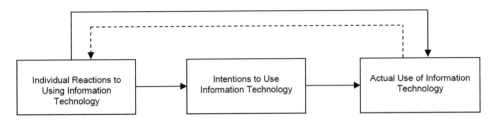

Fig. 5.1 A basic user acceptance model. *Source* Venkatesh et al., 'User Acceptance of Information Technology', 427

impact on 'usage behavior' (UB) (Alghamdi, 2016; Fishbein & Ajzen, 1975; Venkatesh, Thong, & Xu, 2012; Venkatesh, Morris, Davis, & Davis, 2003; Rodrigues, Sarabdeen & Balasubramanian, 2016). In light of the literature, in this research, the behavioral intention (BI) has been used as a predictor of technology acceptance.

5.3 Technology Adoption Models and Theories

In this section, a brief description of various technology adoption theories and models considered for development of UTAUT is given.

5.3.1 Theory of Reasoned Action (TRA)

This theory was proposed by Fishbein and Ajzen in (1975). TRA is the most powerful theory (Marques, Villate & Carvalho, 2011) for predicting the behavior of users in a particular situation. TRA suggests that an important construct of an individual's genuine action is the intention to accomplish that behavior; that is, a function of subjective norms and attitudes toward behavior (Vatanparast, 2010). According to this theory, the behavior of a person is directly influenced by his willingness to adopt or not to adopt that specific behavior (Marques et al., 2011). In TRA theory, the behavior of an individual is measured by his 'intention' to perform a specific task, while the behavioral intention is evaluated by the 'attitude toward behavior' of that person and the

'subjective norm'. In fact, TAM, UTAUT, and many other theories of adoption are driven and heavily influenced by TRA (Vatanparast, 2010). An "*individual's intention is a function of two elementary factors, one personal attitude about behavior and other social influences*" (Ajzen & Fishbein, 1980, p. 6). In other words, TRA contains three main variables: (1) attitude toward the behavior, (2) subjective norms, and (3) behavioral intention. Figure 5.2 shows TRA constructs. The subjective norm can also be stated as the user's perception of the social influence to do that task (Fishbein & Ajzen, 1975; Ajzen & Fishbein, 1980). Attitude toward behavior is defined as a "person's attitude towards performing that behavior" (Kassarjian & Robertson, 1991). According to Fishbein and Ajzen (1975), attitude is like the feelings of a person for performing a specific task that can be assessed through his/her beliefs about the results of that behavior. In spite of its broader acceptance, TRA possesses some limitations. TRA deals with the habitual actions and behaviors that cannot be described by the theory (Al-Qeisi 2009).

5.3.2 Theory of Planned Behavior (TPB)

This theory is an extension of the TRA (Venkatesh & Speier, 1999) with 'perceived behavioral control' as an added variable and it is the user's perception on the difficulty or ease of use. The TPB theory by Ajzen (1991) extended TRA theory by adding perceived behavioral control (Venkatesh & Speier, 1999). Ajzen (1991) demonstrated that perceived behavioral control, subjective norms, and attitudes are

Fig. 5.2 Belief, intention, attitude, and behavior (Fishbein & Ajzen, 1975)

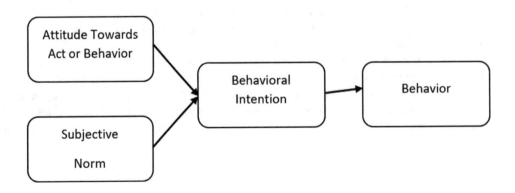

Fig. 5.3 The theory of planned behavior (Ajzen, 1991)

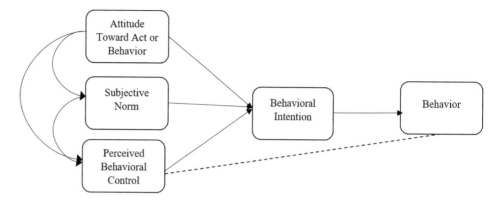

Fig. 5.4 TAM/TPB variables. *Source* Taylor and Todd (1995)

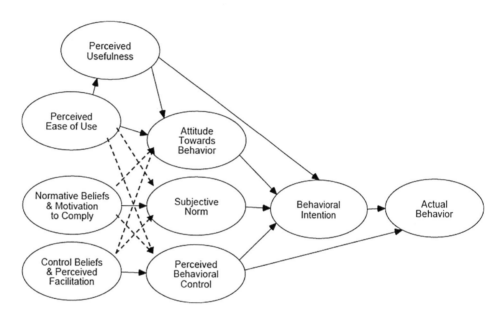

significantly correlated to the intentions and predict the actual behavior of a consumer. As shown in Fig. 5.3, Ajzen (1991) describes three categories of beliefs in TPB: (a) *Behavioral beliefs* that affect behavior and attitude, (b) *Normative beliefs* from an individual's subjective norm, and (c) *Control beliefs* that deliver the basis for perceptions of behavioral control. Figure 5.3 represents the TPB model by Ajzen (1991).

Because of doubts in the definition of perceived behavioral control (PBC), this model has not been verified in empirical settings. Godin and Kok (1996) indicated the drawbacks in the model that some important variables such as personality and demographic variables are also not taken into consideration in TPB.

5.3.3 Combined Model TAM/TPB (TAM/TPB)

A hybrid model was introduced by Taylor and Todd (1995) which combined the factors of the TPB with the 'perceived usefulness' of TAM model. This model offers detailed

variables that help in understanding 'behavioral intentions' and 'usage behavior'. Taylor and Todd (1995) introduced a new variable to the experience of the user on IT and named it 'previous experience'. In general, the combined model reflects that the users' experience levels should also be taken into consideration while studying the theories of acceptance and IT. The TAM/TPB model is shown in Fig. 5.4.

The combined model does not have broader acceptance and is associated with some limitations. For instance, Lim (2003) used this model to investigate "*the adoption of negotiation support systems*" and discovered it to be a valid model, whereas Chau and Hu (2002) found that a combination of these theories was not supportive of their research for technology adoption by individuals in healthcare environments.

5.3.4 Diffusion of Innovation Theory (DOI)

Rogers published his first DOI theory in 1962. Since then, the Rogers DOI framework has been used extensively to discover innovations that have been accepted or rejected. Rogers (1983) defines DOI theory as "*the process by which innovation is communicated through certain channels over time, among the members of a social system*". The diffusion of innovation is the process of gathering information to evaluate technology (Rogers, 1995). The innovation process starts with a basic familiarity of invention that is shaped on an attitude toward it and transitions through to a judgement to either reject or adopt it (Rogers, 1983). Hence, diffusion of technology is a gradual procedure through which new technology is transferred by the associates of a social system through different channels over time (Rogers, 1983). This describes the decision process of acceptance of technology and determines variables that impact the adoption rate of new technology (Marques et al., 2011). Rogers (1983) presented the five attributes that influence the adoption of technology as relative advantage, trialability, compatibility, complexity, and observability, as shown in Fig. 5.5. All the five variables are somewhat interrelated empirically, but all are conceptually separate (Vatanparast, 2010), as shown in Fig. 5.5.

Relative Advantage: It is the "*degree to which innovation is perceived as being better than the idea it supersedes*"; for instance, economic profitability. It is taken as the best predictor of the rate of adoption of technology because relative advantage will reflect the degree to which innovation is better than the older idea (Rogers, 1983). The relative advantage (RA) is similar to perceived usefulness (Nysveen, Pedersen, & Thorbjørnsen, 2005).

Compatibility: It is defined by Rogers (1983) as "*the degree to which an innovation is perceived as consistent with the existing values, past experiences, and needs of potential adopters*". There is a direct positive relation of compatibility on the adoption rate of new technology. Comparing with other attributes, compatibility seems to be comparatively less imperative in predicting the adoption rate, but still, it is very important and an interesting variable in Rogers' theory.

Complexity: As defined by Rogers (1983), complexity is "*the degree to which an innovation is perceived as relatively difficult to understand and use*". The less difficult to understand means that innovation will be less complex and the perceived adoption rate will be higher. In other words, the concept of complexity is negatively associated to its rate of adoption (Rogers 1983).

Trialability: It is defined by Rogers (1983) as "*the degree to which an innovation may be experimented with on a limited basis*". According to Rogers' definition, the trialability of innovation is directly connected to its adoption rate (Roger 1983).

Observability: It is defined by Rogers (1983) as "*the degree to which the results of an innovation are visible to others*". Rogers proposes a positive relationship between the rate of adoption and observability. This means that if the innovation is more visible to the individuals, then the rate of adoption will be faster. Moore and Benbasat (1991) expanded the Rogers model by adding image, demonstrability, and visibility to the model. Moore and Benbasat (1991) borrowed three innovation characteristics (compatibility,

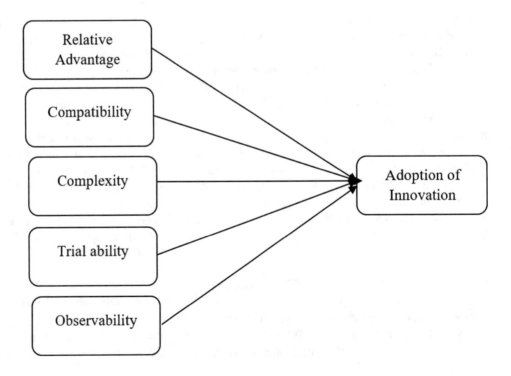

Fig. 5.5 Five different attributes of innovations (Vatanparast, 2010)

relative advantage, and trialability) from Rogers and a fourth characteristic (ease of use) that is also similar to Rogers' complexity.

It is the prediction about the probability of adoption of an innovation. According to Rogers (1983), "*Rate of adoption is the relative speed with which an innovation is adopted by members of a social system. It is generally measured as the number of individuals who adopt a new idea in a specified period, such as a year. So the rate of adoption is a numerical indicator of the steepness of the adoption curve for an innovation*". Based on the unique psychographic characteristics of each group, the adoption/rejection curve by Rogers classifies innovation adopters into five groups as shown in Fig. 5.6. This curve projects the concepts that some people are more inclined to adopt than others.

Figure 5.6 shows the normal frequency distribution that is classified into five classes: (1) innovators, (2) early adopters, (3) early majority, (4) later majority, and (5) laggards.

(1) **Innovators**: These include the 'techies' and the 'experimenters' who follow new invention as soon as it is available. The population of this category is ± 2.5% of the adopter.

(2) **Early adopters** are generally not technologists, although they exploit the new technology. Early adopters are the 'visionaries' who are interested in new technology to explore it for solving technical problems and tasks. The ± 13.5% of technology adopters are popular and are social leaders.

(3) **Early majority** are the 'pragmatists' who are generally comfortable with a new invention. The early majority concentrate on technical problems than on the available tools to address the problems. The next 34% of adopters are slow and cautious about the adoption of innovation.

(4) **Late majority** are the conservatives or 'skeptics' who are less comfortable with technology. The next 34% of adopters are skeptical, uncertain, traditional, and have lesser socio-economic standing.

(5) **Laggards**: Mostly they are not interested in new technology and will never adopt it. However, generally, they buy and use technology when it is concealed with other products.

One of the most significant contributions of this model is the definition of innovation in the decision process, which starts with the individual's knowledge (awareness) about new technology and finishes with the confirmation of the rejection or adoption of that innovation.

5.3.5 Social Cognitive Theory (SCT)

This theory supports the answers to questions such as: Are work and other life roles assumed as more or less relevant? And how can individuals take self-directivity in its development progress (Marques et al., 2011)? SCT theory offers a foundation for knowing, predicting, and modifying human behavior accordingly. The theory classifies the behavior of humans as an interaction of personal variables, the environment, and behavior (Bandura 1986). SCT theory is used to explain how people attain and continue specific behavioral patterns and delivers the foundation for intervention strategies (Bandura, 1986). Bandura's SCT theory highlights the importance of the Lent, Brown, and Hackett (1994) assumption that there is a complex set of variables such as gender, culture, state of health, and socio-structure that work together and influence the cognitions. SCT theory describes that understanding and learning develop due to the interaction among three variables (as shown in Fig. 5.7): behavior, personal variables, and environmental variables (Pajares & Schunk, 2002).

This theory deals with constructs like self-efficacy, anxiety, and effects in defining usage behavior (Bandura, 1986; Venkatesh et al., 2003).

5.3.6 Motivation Model (MM)

Deci and Ryan (1987) proposed the model that "*self-determination is a human quality that involves the experience of choice, having choices and making choices*" (Al-Qeisi, 2009, p. 68). The theory of motivation deals with the concept that the behavior is found by both extrinsic and intrinsic motivations. *Intrinsic motivation* is the satisfaction achieved by acting itself (Vroom, 1964). It is the pleasure associated with the performance of an activity (Bagozzi, Davis, & Warshaw, 1992). *Extrinsic motivation* considers an action due to a reward, such as increased performance (Deci, 1975). It refers to the result of an activity and the value of achieving it (Marques et al., 2011). In Fig. 5.8, Igbaria, Parasuraman

Fig. 5.6 Diffusion of innovation adopters (Rogers, 1983)

Fig. 5.7 *Source* Social cognitive theory (Pajares & Schunk, 2002)

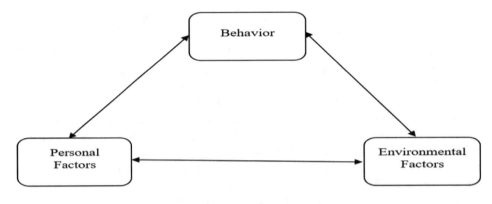

Fig. 5.8 A motivational model. *Source* Igbaria et al. (1996): A motivational model

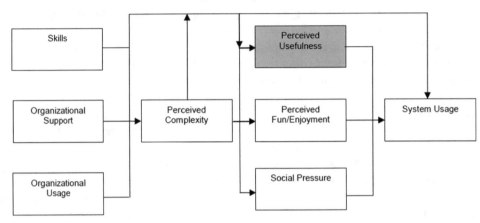

and Baroudi (1996) present a motivational model specific to microcomputer usage (p. 135).

5.3.7 Model of PC Utilization (MPCU)

Thompson, Higgins and Howell (1991) developed the model of PC utilization to describe the problems of PC utilization (Jen, Lu & Liu, 2009). The MPCU was driven by the theory of human behavior established by Triandis (1977). The key variables included in MPCU are social, facilitating conditions, and long-term consequences (Jen et al., 2009). This model delivers a basic concept on how behaviors are developed under the influence of different variables. In addition to predicting the intention to use, many researchers also used the MPCU concept to predict the use of personal computers (Triandis, 1977). Thus, this model has been used as a predictor of technology acceptance in various areas of IT (Marques et al., 2011) (Fig. 5.9).

5.3.8 Technology Acceptance Model (TAM, TAM2, and TAM3)

Technology adoption model (TAM) by Davis (1989) is derived from the TRA and was originally designed for the

area of information technology (Marques et al., 2011). The key goal of TAM "*is to explain the determinants of computer acceptance that is general and capable of explaining user behavior across a broad range of end-user computing technologies and user populations...*" (Davis et al., 1989, p. 985). According to TAM, the willingness of a person to use a particular system in the future is the 'behavioral intention' of that person to use that system, and it is based on two variables: 'perceived ease of use' and 'perceived usefulness' (Davis et al., 1989; Shiau & Chau, 2015). TAM has been successfully tested by various researchers (Legris et al., 2003) not only on a theoretical but also on an empirical basis (Hu, Chau, Sheng, & Tam, 1999). **Strengths of TAM**: The TAM model has been supported by extensive empirical research over the years (Davis and Venkatesh, 1996; Chau and Hu, 2002; Legris, Ingham & Collerette, 2003), and this empirical support is the key strength of the TAM model. The prior research shows that the TAM model is the most extensively accepted model in the research of information technology (Davis, 1989; Jen et al., 2009; Marques et al., 2011; Shiau & Chau, 2015). TAM is considered as the most robust technology adoption model (Baptista & Oliveira, 2015), and it has become one of the primary adoption theories applied to higher education that provides a solid theoretical foundation to investigate the intention of technology adoption. The TAM model could predict the technology

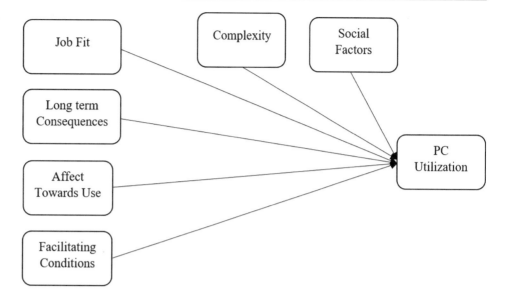

Fig. 5.9 The model of personal computer utilization (MPCU). *Source* Thompson and Higgins (1991)

acceptance in roughly 40% of the situations (Taylor & Todd, 1995; Venkatesh & Davis, 2000). The empirical testing of TAM has also encouraged the researchers to extend the model having a variety of the constructs, particularly 'external constructs', in this model. TAM has become one of the fundamental and most adopted theoretical models in the field of information security (Lee, 2004). It has provided researchers with "valid, reliable, and easy to administer scales for the key constructs" (Venkatesh, Davis, & Morris, 2007, p. 268). The TAM assists managers in explaining how they can get employees to accept new technologies (Pearlson & Sauders, 2003).

Limitations of TAM: TAM is limited to two determinants: perceived usefulness and perceived ease of use in foreseeing user behavior and fails to identify other adoptance determinants, as demonstrated by future models such as TAM2, TAM3, and UTAUT. Another constraint of this model is that TAM does not consider system and organization variables such as system characteristics, training, the financial cost to the individual, management, and technical support (Handy, Whiddett, & Hunter, 2001). Van Biljon (2006) criticized the TAM model for failing to cover cultural and social variables. Another key criticism is that the TAM model does not recognize individual differences such as gender, age, and experience which may impact a user's attitude toward adoption of that particular system (Agarwal & Prasad, 1999). An important deficiency, pointed out by Davis et al. (1989), is that a subjective norm (SN) or social influence is lacking from the TAM model. Dishaw and Strong (1999) pointed out TAM's deficiency in that TAM lacks the inclusion of external variables. It ignores the most important barriers such as cost, time, and lack of expertise. TAM has failed to supply information regarding user adoption of a specific technology

due to its generality; also, TAM neglects many important sources of variance (Mathieson, Peacock, & Chin, 2001).

Major Constructs of TAM: TAM adapts the 'belief → attitude → intention → behavior' relationship from TRA. TAM suggests that two concepts, perceived ease of use (PE) and perceived usefulness (PU), are the main elements for acceptance of IT. Figure 5.10 depicts the TAM model and its constructs by Davis (1989).

- **Perceived Usefulness (PU)**: It is the degree to which an individual considers that utilizing a specific system would improve his/her performance (Davis, 1989). PU is a significant determinant in Davis's TAM model that explains the use of technology by concentrating on the users' attitudes toward the technology, and their subsequent intention to use it. Perceived usefulness has various dimensions, such as effectiveness, usefulness for job, and efficiency.

- **Perceived Ease of Use (PEU)** is the *"degree to which a person believes that using a particular system would be free of effort"* (Davis, 1989: 320). PEU is related to the ease of learning, that is, physical as well as mental efforts of using the technology. The study shows that PEU influences attitude, intentions, and behavior. PEU also influences the PU because if the technology is easy to use then usefulness will be expected to be increased. Previous research shows that PU is considered to be a more important determinant than PEU, which is the determinant of PU.

5.3.9 Extended TAM (TAM 2)

To overcome the lack in TAM, numerous approaches to improving TAM have been suggested (Van Biljon & Kotze,

Fig. 5.10 Technology acceptance model (Davis, 1989)

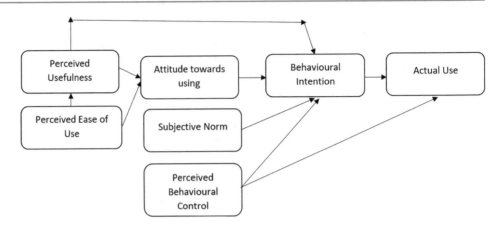

2007), such as TAM2 and TAM3. Compeau and Higgins (1995) also suggested extending TAM by including computer self-efficacy. To address the weakness and limitations of TAM, Venkatesh and Davis (2000) extended the TAM model to incorporate: (a) *social influence concepts,* for instance, subjective norm and voluntariness, and (b) *cognitive instrumental concepts,* for instance, perceived ease of use, job relevance, and output quality. "The TAM2 model introduced a social dimension to capture the influence of the end-user environment" (Hadji & Degoulet, 2016). In TAM2, the variables of perceived usefulness are perceived ease of use, job relevance, subjective norm, image, result demonstrability, and output quality. Experience and voluntariness are two identified moderators. The TAM2 model suggests that perceived ease of use, usefulness, and subjective norms are significant variables of usage intentions (Vatanparast, 2010). TAM explains approximately 40% of behavioral intention, but TAM2 explains the behavioral intention of 10–20% more of the variance than TAM (Davis et al., 1989) (Fig. 5.11).

5.3.10 TAM3

Venkatesh and Bala (2008) improved the TAM and TAM2 models to develop an integrated and comprehensive model.

The new model is called TAM3, which pinpoints the various variables of acceptance and the use of new technology, which were adopted from previous research on technology adoption. With three extensions to TAM2, the constructs of TAM2 for PU were joined with PEOU (Venkatesh & Davis, 2000) to develop the integrated model, that is, TAM3. This new model (TAM3) was almost similar to TAM2, except that it included an *anchor* and *adjustment* as external variables to describe *perceived ease of use*. The category anchors included *computer self-efficacy, perceptions of external control, computer playfulness*, and *computer anxiety*, whereas the category adjustment included *objective usability* and *perceived enjoyment* (Venkatesh & Bala, 2008). Another difference was that the experience was the moderator between two of the anchor factors, computer anxiety and computer, as well as two adjustment variables, that is objective usability and perceived enjoyment. Although TAM3 is an enhanced version of previous models, it is not considered in this study due to the following drawbacks: TAM3 theory is a new theory and not enough research has been published validating it. In TAM3 research, all participants were drawn from financial services companies, accounting service companies, manufacturing companies, and international investment banking firms. All of the contributors were staffs of non-academic organizations and no participants were taken from educational institutions.

5.4 Unified Theory of Acceptance and Use of Technology (UTAUT)

Venkatesh et al. (2003) compared the conceptual and empirical similarities of eight models of technology adoption and developed a model that combines the most common variables of all eight models (Diep, Cocquyt, Zhu, & Vanwing, 2016; Marques, Villate, & Carvalho, 2011; Mohammadyari & Singh, 2015; Venkatesh, Morris, Davis, & Davis, 2003). The new model is referred to as a 'unified theory of acceptance and use of technology' (De Wit, Heerwegh &

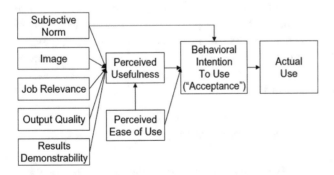

Fig. 5.11 Extension of the TAM–TAM2 (Venkatesh & Davis, 2000). *Source* Venkatesh and Davis, 'A Theoretical Extension', 188

Fig. 5.12 UTAUT model by Venkatesh et al. (2003)

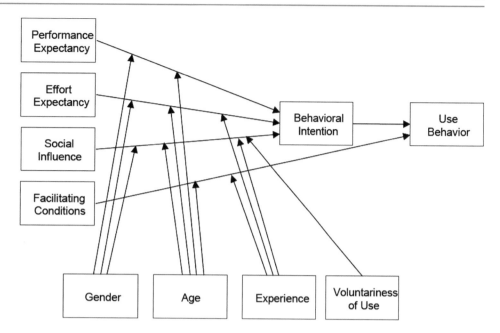

Verhoeven, 2014). It provides a greater understanding of the acceptance of technology and it is a widely used model in the field of ICT acceptance (Rodrigues et al., 2016). Various researchers acknowledged that UTAUT provides a better understanding regarding an individual's behavior toward acceptance of innovation than other similar models and theories (Baptista & Oliveira, 2015; Gilbert, Balestrini & Littleboy, 2004; Schaper & Pervan, 2007; Venkatesh et al., 2003; Wu, Hsu, & Hwang, 2007). The selection of using a unified model was inspired by its excellent explanatory and comprehensiveness power than other models of technology acceptance (Kripanont, 2007; Tibenderana & Ogao, 2009; Venkatesh et al., 2003). Various studies have validated UTAUT in various environments (Lakhal, Khechine, & Pascot, 2013), such as education, banking (AbuShanab, Pearson, & Setterstrom, 2010; Tibenderana & Ogao, 2009; Williams, 2009), organizations (Al-Gahtani, et al., 2007; Bourbon, & Hollet-Haudebert, 2009; Eckhardt, Laumer, & Weitzel, 2009), and tourism (San Martin, & Herrero, 2012). It has become a prominent model and has been cited more than 5000 times by Google Scholars (Rodrigues et al., 2016). The results of empirical research by Venkatesh et al. (2003) explained a greater percentage (70%) of the variance of 'intention to use' and 'usage behavior' than any other model (Al-Gahtani, Hubona & Wang, 2007; Rodrigues et al., 2016). It has become a benchmark in IS acceptance (Rodrigues et al., 2016). This indicates that this model is comprehensive and offers an excellent framework for explaining the usage and acceptance behaviors of new technology such as LMS. However, there some limitations associated with UTAUT. Comparing with other models such as TAM or DOI, UTAUT is a comparatively new model and requires further research study to validate its constructs,

replicate findings, and validate its strength (Straub, 2009). Although UTAUT has been tested in various information system (IS) research, there are still limitations (Negahban, & Chung, 2014) and areas open for further research in technological areas (Baptista & Oliveira, 2015) that might fall within the 30% unexplained adoption of this model (Baron, Patterson, & Harris, 2006). In addition, this model does not contain individual variables such as self-motivation, perceived playfulness that may help explain IS acceptance and its usage.

As shown in Fig. 5.12, the original UTAUT model offered seven constructs linked to technology adoption: (1) performance expectancy, (2) effort expectancy, (3) facilitating conditions, (4) social influence, (5) computer anxiety, (6) computer self-efficacy, and (7) attitude toward technology usage. The age, gender, experience, and voluntariness of use were the moderating variables of the UTAUT model. During its empirical validation, Venkatesh et al. (2003) excluded many variables that were less significant in prior research. The variables such as anxiety, self-efficacy, and attitude toward using technology were three variables found to be not determinants of *behavioral intention (BI)* and were removed from the UTAUT model (Venkatesh et al., 2003). Figure 5.12 shows the resultant UTAUT model.

5.4.1   Key Constructs of UTAUT

The UTAUT model is composed of two dependent variables 'behavioral intention' and 'usage behavior', and four independent variables: 'performance expectancy', called perceived usefulness, 'effort expectancy', called perceived ease of use, 'facilitating conditions' and 'social influence', which have a direct influence on the intention to use the system. However, age, experience, gender, and voluntariness act as moderating variables (Venkatesh et al., 2003). This means that if the values of these variables are higher, the value of behavioral intention to use the technology is higher, and so is the individual's acceptance of the technology (Venkatesh et al., 2003).

- **Performance Expectancy (PE)**: Performance expectancy is the degree to which a user perceives that using technology will assist him or her to achieve benefits from job performance. Venkatesh et al. (2003) stated that *"performance expectancy is the strongest predictor of intention"* (p. 447). It is an independent variable of the UTAUT model and is the most significant determinant of an individual's behavioral intentions to use a technology (Al-Gahtani et al., 2007; Al-sobhi, Weerakkody, & El-haddadeh, 2011; AlAwadhi & Morris, 2008; Chen, Lai, & Ho, 2015; Venkatesh et al., 2003). Bandyopadhyay and Fraccastoro (2007) found that performance expectancy (PE) is strongly related to behavioral intention (BI) among consumer prepayment metering systems (p. 535). With the use of Internet banking software in Jordan, AbuShanab et al. (2010) determined that performance expectancy (PE) is strongly associated with the behavioral intention (BI) among bank customers (p. 511). The performance expectancy (PE) in UTAUT, perceived usefulness in (TAM/TAM2 and C-TAM-TPB), extrinsic motivation (MM), outcome expectations (SCT), relative advantages in DOI, or job-fit (MPCU) are important constructs in measuring the advantages achieved by using the technology (Shareef et al., 2011).
- **Effort Expectancy (EE)**: Venkatesh et al. (2003) defined that effort expectancy is "the degree of ease associated with the use of the system" (Venkatesh et al., 2003, p. 450). Effort expectancy refers to the ease of use a user linked with the use of technology, as perceived by the user. Venkatesh and Zhang (2010) emphasized that effort expectancy (EE) is a very strong predictor of technology adoption. Rogers (2003) commented that if a technology is perceived to be difficult to understand or to use, then it would be regarded as complexity. In the UTAUT model, Venkatesh et al. (2003) considered the equivalent factors from other models that capture the concept of EE are complexity (DOI, MPCU) and perceived ease of use (TAM/TAM2).
- **Social Influence (SI)**: Social influence (SI) includes the social pressure exercised on a person by the beliefs of other individuals or groups. The social influence is "the degree to which an individual perceives that important others believe he or she should use the new system"

(Venkatesh et al., 2003, p. 451). This determinant is based on the supposition that user behavior is influenced by his/her perception of how his/her usage of technology is viewed by other people (Venkatesh et al., 2012). According to Rogers (2003), the decisions of adoption are socially influenced by the role available to a person or a group of people. Empirical outcomes show that social influence employs a positive influence on intention to use the technology (Venkatesh & Davis, 2000; Venkatesh, et al., 2003; Wong, Teo, & Russo, 2012). The construct SI is known as the *subjective norm* in other models such as TAM2, TRA, TAM/TPB, and TPB/DTPB models. It is known as *social factors* in the MPCU model and *image* in the DOI model (Yeow & Loo, 2009). Based on the concepts related to social influence defined by authors such as Ajzen (1991), Davis et al., (1989), Fishbein and Ajzen, (1975), and Taylor and Todd, (1995), the SI is categorized into two sub-variables: peer social influence and general social influence. The general social influence, in general, means what other people think of the use of technologies and the support offered by the management about the use of the technology. Peer social influence was developed from colleagues or peers who used the technology. Previous research on attitudes and intention toward technology have revealed that the influence of society is an important and critical aspect that influences personal belief to make decisions about technology adoption (Anderson, Al-Gahtani, & Hubona, 2011). Venkatesh et al. (2003) suggest a positive and direct link between social influence (both peer social and general influence) and intentional behavior to use technology.

- **Facilitating Conditions (FC)**: Venkatesh et al. (2003) defined that facilitating conditions are "the degree to which an individual believes that an organizational and technical infrastructure exists to support the use of the system" (Venkatesh et al., 2003, p. 453). In the UTAUT model, Venkatesh et al. (2003) considered other models having constructs similar to facilitating conditions. The researchers found that FC has a direct impact on the actual use of technology (Venkatesh et al., 2003). There is somewhat of a contradiction in previous studies concerning the relationship between 'actual use' of technology and facilitating conditions. Some authors argue that facilitating conditions are directly linked with behavioral intentions (Eckhardt et al., 2009; San Martín, & Herrero, 2012). On the other hand, some authors excluded FC from their research study (Bandyopadhyay & Fraccastoro 2007). According to Wu, Tao, & Yang (2007), the facilitating conditions have a strong effect on the intention to adopt the technology. The equivalent variables from other models having a similar concept of facilitating conditions are 'compatibility' from the DOI model and perceived behavioral control from TPB/DTPB, TAM/TPB, and MPCU (Yeow & Loo, 2009).

5.5 UTAUT2: An Extension of the UTAUT Model

In 2012, Venkatesh, Thong, and Xu revised and updated UTAUT and included various constructs such as habit, hedonic motivation, and price value. The new model is called the modified (or extended) unified theory of acceptance and use of technology (UTAUT2). The validation of the study of UTAUT2 included two-stage online survey results of 1512 mobile consumers in Hong Kong in the context of the adoption of mobile phones (Venkatesh et al., 2012, p. 166).

Constructs of UTAUT2: The UTAUT2 (Venkatesh et al., 2012) model has two dependent variables: 'behavioral intention' and 'usage behavior'. The independent variables of UTAUT2 include: 'effort expectancy', 'performance expectancy', 'social influence', 'facilitating conditions', 'hedonic motivation', 'habit', and 'price value'. The gender, age, and experience act as moderating variables of UTAUT2 (Venkatesh et al., 2003). Venkatesh et al. (2012) added three new variables (hedonic motivation, habit, and price value) with UTAUT2 in the context of mobile phones.

- **Hedonic Motivation (HM)**: According to Venkatesh et al. (2012) hedonic motivation is "the fun or pleasure derived from using a technology" (p. 161). According to Brown and Venkatesh (2005), as cited by Venkatesh et al. (2012), hedonic motivation "plays an important role in determining technology acceptance and use" (p. 161) and it is a significant predictor of technology adoption (Brown & Venkatesh, 2005; Van der Heijden, 2004). In IS research, the hedonic motivation, also known as perceived enjoyment, is the key variable that impacts an individual to adapt and use the technology (Thong et al., 2006).
- **Habit**: Habit is the automatic behavior that enables learning on how to use the technology. The previous use of technology becomes a habit, and habit becomes a significant determinant of future use (Kim & Malhotra, 2005). Venkatesh et al. (2012) outlined the difference between habit and previous experience. According to

Venkatesh et al. (2012), the 'experience' refers to an opportunity to use new technology and is normally operationalized gradually from the initial use of a technology by a user, whereas 'habit' is related to the automatic behavior of people because of the learning. Experience is associated with the chronological time that results in the formation of different levels of habit. For instance, over three months, different people will have different levels of habit depending upon the use of technology (Venkatesh et al., 2012). Limayem, Hirt and Cheung (2007) stated that habit is not only linked with the usage of technology, but it also moderates the impact of intention on the usage of technology such that behavioral intention becomes less important with the increase of habit. The empirical findings show that habit influences technology use (Venkatesh et al., 2012).

- **Price Value**: The cost or price may be the key influence of a user's technology adoption and use (Venkatesh et al., 2012). The pricing and cost issues may have an impact on the consumers' technology use (Abdullah & Khanam, 2016), but this study deals with the LMS adoption by the instructors. The cost or price structure of the LMS is relevant to higher management or with the organizational settings; the employees (instructors) do not bear the monetary cost (Venkatesh et al., 2012) of LMS. Therefore, the construct 'Price Value' has been excluded from this research.

Figure 5.13 represents the original UTAUT2 model. UTAUT2 (Venkatesh et al., 2012) is different from the original UTAUT (Venkatesh et al., 2003) model because of the models such as the MPCU, UTAUT, TAM, TAM2, and TAM3 were formulated in an organizational and business context, whereas UTAUT2 was developed to predict the *user behavior* of consumers in the context of mobile phones. In UTAUT2, the moderating variable, 'voluntariness' between social influence and behavioral intention has been removed. In UTAUT2, three new variables, that is, price value, habit, and hedonic motivation have been added as independent with the behavioral intention (BI).

5.6 Variables Considered for UTAUT and UTAUT2

It is evident from Table 5.1 that UTAUT and UTAUT 2 included almost all variables used in other models (San Martin & Herrero, 2012).

Fig. 5.13 UTAUT2 model. *Source* Venkatesh et al. (2012)

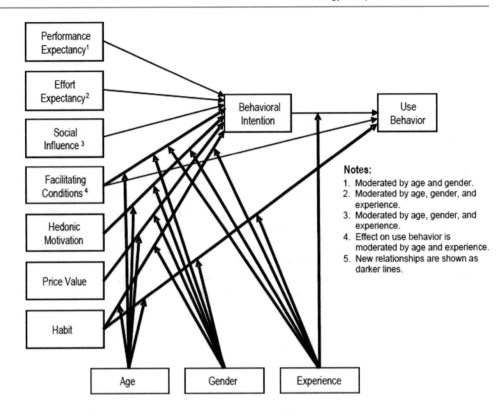

Table 5.1 Summary of dependent and independent variables of technology adoption models

Theory/Models	Author(s)/Date	Dependent variable	Independent variable
Theory of research action (TRA)	Fishbein and Ajzen (1975)	Behavioral intention, behavior	Attitude toward behavior, subjective norms
Theory of planned behavior (TPB)	Schifter and Ajzen (1985), Ajzen (1991)	Behavioral intention, behavior	Attitude toward behavior, subjective norm, perceived behavioral control
Decomposed theory of planned behavior (DTPB)	Taylor and Todd (1995)	Behavioral intention, behavior	Attitude toward behavior, subjective norm, perceived behavioral control, perceived usefulness
Diffusion of innovation (DOI)/innovation diffusion theory (IDT)	Rogers (1983),/ Moore and Benbasat (1991)	Adoption of innovation or implementation success	Ease of use, relative advantage, image, compatibility, visibility, voluntariness of use, results demonstrability
Socio-cognitive theory (SCT)	Compeau and Higgins (1995)	Usage behavior	Outcome expectations (performance, personal), self-efficacy, affect, anxiety
Technology acceptance model (TAM)	Davis (1989), Davis et al. (1989)	Behavioral intention to use, system usage	Perceived usefulness, perceived ease of use, subjective norm
Model of PC utilization (MPCU)	Thompson and Higgins (1991)	PC usage	Job-fit, complexity, affect toward use, social factors, long-term consequences, facilitating conditions
Motivation model (MM)	Davis and Warshaw (1992)	Usage of technology	Extrinsic motivation, Intrinsic motivation
UTAUT	Venkatesh et al. (2003)	Behavioral intention, Usage behavior	Effort expectancy, performance expectancy, facilitating conditions, social influence, gender, experience, age voluntariness of use (moderators)
UTAUT2	Venkatesh et al. (2012)	Behavioral intention, usage behavior	Effort expectancy, social influence, facilitating conditions, hedonic, motivation, habit, price value, gender, age, experience (moderators)

Source Adapted from San Martin and Herrero (2012)

5.7 Extension of the UTAUT2 Model

Many researchers have argued that the UTAUT model ignored the variables that might be significant predictors of technology acceptance and usage (Yeow & Loo, 2009). Therefore, to capture the ignored variables, many critics have highlighted the importance of the extension of this model (Alazzam et al., 2016; Berthon, Pitt, Ewing, & Carr, 2002; Venkatesh et al., 2003). Venkatesh et al. (2003) and Venkatesh, Speier and Morris (2002) suggested that the future model should be supported with enough constructs such as individual constructs and technology fit. Dwivedi, Rana, Chen and Williams (2011) pointed out that the trend of using external variables with UTAUT is increasing. In a study of '*UTAUT external variables*', Dwivedi et al. (2011) found that 22 out of 43 researchers have used external constructs in their research and the remaining 21 used the original variables of UTAUT in their studies. Venkatesh et al. (2012) emphasized the need for extension of the UTAUT/UTAUT2 models and stated that the extension of UTAUT/UTAUT2 is valuable in understanding and expanding technology acceptance boundaries (Venkatesh et al., 2012, p. 158) as well as expanding the scope and generalizability of the model (Venkatesh et al., 2012, p. 160). Three types of extensions are suggested by Venkatesh et al. (2012). The suggested extension of the model is a new technological extension, such as the health information system (Chang, Hwang, Hung & Li, 2007) or new user populations such as educational users, consumers, and new cultural settings (Gupta et al., 2008). Another type of extension is the inclusion of new constructs within UTAUT (Sun, Bhattacherjee, & Ma, 2009). The third type of extension is the addition of exogenous predictors of the UTAUT variables (Yi et al., 2006).

In the guidance of the literature review, the technological extension includes the use of LMS technology by instructors in the context of higher educational institutions (HEIs). Another extension of the model was the inclusion of moderating variables (Al-Gahtani et al., 2007; Kripanont, 2007; Tibenderana & Ogao, 2009; Venkatesh et al., 2012). Table 5.2 provides a list of external variables used by different researchers. It shows that anxiety, trust, attitude, self-efficacy, PU, PEOU, perceived credibility, and perceived risk are the most common external variables.

5.8 Moderating Variables (Age, Experience, Gender, Technology Awareness, and Culture)

A moderator is a quantitative or qualitative variable that influences the direction of the relationship and strength between two variables (Baron & Kenny, 1986; Kripanont, 2007; Lakhal et al., 2013; Schaper & Pervan, 2007; Serenko,

Turel, & Yol, 2006; Venkatesh et al., 2003). Literature shows that demographic features such as age, gender, marital status, and family structure influence the acceptance of technology and therefore cannot be neglected. The concept of moderators and core constructs is well documented in their research by several researchers (Kripanont, 2007; Schaper & Pervan, 2007; Venkatesh et al., 2003).

The moderators of UTAUT are gender, age, experience, and voluntariness (Venkatesh et al., 2003), whereas those of the UTAUT2 model are: age, gender, and experience (Baptista & Oliveira, 2015; Venkatesh et al., 2012). The moderators of this study include *age, experience, technology awareness, racial demography,* and *cultural dimensions*.

5.8.1 Age (As a Moderating Variable)

Age is the key personal characteristic included in demographic variables. Research shows that the adoption of technology is associated with age (Venkatesh et al. 2003; Morris & Venkatesh, 2000). It was found that age moderates the influence of all of the variables on behavioral intentions (BI). Venkatesh et al. (2003) stated that age and gender play a moderating role in UTAUT model for BI to adopt a technology. Venkatesh et al. (2003) found that age is the moderator of adoption for EE, PE, and SI in the UTAUT model. Bandyopadhyay and Fraccastoro (2007) stated that age and gender are the moderators between PE, EE, and SI, and behavioral intentions to use the technology. According to Morris and Venkatesh (2000), age has a moderating impact on PE such that it has a greater impact on younger men than in older. However, Cheng et al. (2012) reported that age and gender mediate the relationship between SI and BI. Research by Al-Gahtani (2003) in the Saudi context revealed more chances of adoption of technology in older men, whereas Oyelaran-Oyeyinka and Adeya (2003) identified that younger men and women are more likely to use innovation. Research reveals that older people have greater difficulties in adopting new technology than younger people (Ellis & Allaire, 1999). Previous research shows that variables connected to effort expectancy (EE) are good predictors of behavioral intention for females (Venkatesh & Morris, 2000) and older workers (Morris & Venkatesh, 2000). Facilitating conditions are moderated by experience and age (Venkatesh et al., 2003). Similarly, AbuShanab et al. (2009) found that both age and gender significantly moderate both performance expectancy (PE) and effort expectancy (EE).

5.8.2 Experience (As a Moderating Variable)

Experience is a key personal characteristic recognized as a moderating variable in UTAUT2. Prior research shows

	External variables used with UTAUT/UTAUT2 model	Author (year)
Table 5.2 External variables used with the UTAUT/UTAUT2 models	Trust in ERP context	Alazzam et al. (2016)
	Confidentiality and trust using UTAUT	Rodrigues et al. (2016)
	In UTAUT2, cultural moderators were integrated to evaluate the impact of culture in the mobile banking context	Baptista and Oliveira (2015)
	UTAUT2 in the healthcare context	Slade and Williams (2013)
	Autonomy, peer influence	Lakhal, Khechine and Pascot (2013)
	Self-efficacy, social norms, behavior control	Giannakos and Vlamos (2013)
	Time of involvement, frequency of use	Hsu (2012)
	Attitude toward behavior	Ajzen (1991)
	Attitude, self-efficacy, awareness, position, user-demography (age, gender)	Dulle and Minishi-Majanja (2011)
	Attitude, self-efficacy	Dulle and Minishi-Majanja (2011)
	Self-efficacy, social norms	Alice (2011)
	Compatibility, behavior control, access	Islam (2011)
	Social norms, trust, awareness	Gholami, Ogun, Koh and Lim (2010)
	Trust, permission (autonomy), utility expectancy, entertainment, user-demography (age, gender, education)	Zolfaghar, Khoshalhan and Rabiei (2010)
	Self-efficacy, anxiety, perceived credibility, attitude	Yeow, Yuen, Tong and Lim (2008)
	Task technology fit	Zhou, Lu and Wang (2010)
	Self-efficacy, the voluntariness of use, anxiety	Curtis et al. (2010)
	Self-efficacy, trust, belief, perceived risk, the disposition to trust	Luo, Chea and Chen (2011)
	Optimism bias, perceived risk, the trust of the e-file system	Schaupp, Carter and McBride (2010)
	Objective norm, subjective norm	Laumer, Eckhardt and Weitzel (2010)
	Trust, past transactions, internet, experience, gender, age	Chiu et al. (2010)
	Anxiety, self-efficacy, trust, perceived risk, locus of control, personal innovativeness	Abu-Shanab and Pearson (2009)
	Perceived credibility, anxiety	Yeow and Loo (2009)
	Ease of use (TAM), experience	Novakovic, McGill and Dixon (2009)
	Attitude, anxiety, self-efficacy	Jong and Wang (2009)
	Voluntariness, experience, knowledge	Kijsanayotin, Pannarunothai and Speedie (2009)
	Self-efficacy, perceived credibility, anxiety, attitude	YenYuen and Yeow (2009)
	Anxiety, perceived credibility	Loo, Yeow and Chong (2009)
	Computer self-efficacy, result demonstrability, resistance to change, computer anxiety, relevance, screen design, terminology	Nov and Ye (2009)
	Self-efficacy, trust, perceived security	Shin (2009)
	Self-efficacy, attitude, anxiety, perceived usefulness, ease of use, training	Aggelidis and Chatzoglou (2009)
	PU, PEOU, cultural influence, and human nature influence, socio-economic factors, demographic factors, and personal factors	Van Biljon and Kotzé (2008)
	Age, gender, digital media preference, societal position, educational level, digital media experience, digital media access, attitude toward use, family position, knowledge of services	Van Dijk, Peters and Ebbers (2008)
	Computer self-efficacy, breadth of use, social influences, satisfaction, relative advantage, risk aversion, perceived security, risk aversion	Ye, Seo, Desouza, Sangareddy and Jha (2008)
	Online social support, online support expectancy	Lin and Anol (2008)
	Computer self-efficacy, utility, attainment value, intrinsic value (playfulness), anxiety, social isolation, delay in responses, anxiety, risk of arbitrary learning	Chiu and Wang (2008)

(continued)

Table 5.2 (continued)

External variables used with UTAUT/UTAUT2 model	Author (year)
Individual innovativeness, task technology fit, compatibility	He and Lu (2007)
Compatibility, computer anxiety, computer attitude, acceptance motivation, organizational facilitation	Dadayan and Ferro (2005)

Source Adapted from Dwivedi et al. (2011)

varying results about the experience as a moderator. The results of various researchers show that experience moderates the relationship of most constructs to behavioral intentions (BI). Experience also moderates the relationship of habit (H) facilitating conditions (FC) and behavioral intention (BI) as a direct determinant of use behavior (UB). Age, gender, and experience also have a combined effect on the correlation between facilitating conditions (FC) and behavioral intention BI. Hall and Mansfield (1975) discovered that gender, age, and experience are the moderates of technology acceptance. Furthermore, according to Venkatesh et al. (2012), with the joint impact of gender and age, experience further moderates the association between behavioral intention (BI) and facilitating conditions (FC). The relationship between BI and FC is moderated by experience (Venkatesh et al., 2012). Hence, having more experience with technology results in more familiarity, confidence, and understanding with the technology, therefore reducing the dependency on facilitating conditions (Alba & Hutchinson, 1987).

5.8.3 Gender (As a Moderating Variable)

Gender is a moderator that affects all constructs of behavioral intentions (BI). Venkatesh et al. (2003) claimed that gender moderates the association between a) EE and BI, (b) PE and BI, and (c) SI and BI. Venkatesh et al. (2003) found that gender possesses a moderating effect on the social norm (Venkatesh et al., 2003). Women are found to be more conscious of SI than men and hence the impact of SI on intentions was greater for women, especially for older women. Venkatesh and Morris (2000) reported that the technology adoption decision is strongly influenced by PE for men and by EE and SI for women. *The gender is not considered for this study because the population of this study in the HEIs of Saudi Arabia is male.*

5.8.4 Technology Awareness (As a Moderating Variable)

The literature shows a significant relationship between awareness and behavioral intension (Faruq & Ahmad, 2013) to adopt new technology. Bardram and Hansen (2010) found

a strong relationship between technological awareness and its use. Charbaji and Mikdashi (2003) used awareness in his study and found that awareness strongly influences the BI to use e-government. The new technology may be adopted by the people if the people are aware of the technology to an adequate level (Lee & Wu, 2011). Similarly, Rehman, Esichaikul and Kamal (2012) argue that the adoption of technological services requires that individuals should be aware of services. The adoption of new technology by end-users is influenced by their personal beliefs and attitudes that have been highlighted in many theories and models such as TRA, TPB, and TAM. The personal belief and attitudes of the individual are more likely to be established if the individuals are aware of new technology when initially launched. The 'awareness', according to Sun and Fang (2016) is, "The degree to which a person thinks about how the technology fits the individual's local specifics and his/her own needs". The technology is designed for specific tasks that work under specific technical environments such as learning ability and availability of technical support (Burton-Jones & Grange, 2013). To achieve a better benefit of technology, the users of the technology need to be aware of the issues associated with the technology (Burton-Jones & Grange, 2013; Porter & Graham, 2016; Sun & Fang, 2016). Adoption of technology with a less mindful adoption decision may lead to the wastage of resources and investment due to lack of alignments between the context and the technology. The meaning of the technology contexts is that technology users are aware of the advantages and disadvantages of the technology (Sun & Fang, 2016). The awareness of the advantages and disadvantages of technology builds confidence in using it (Ahmed et al., 2016), although the real challenge is the effective adoption of educational technology (Ismail, 2016). The lack of technology awareness has been cited as the 'first barrier' to the adoption of innovative technologies (Riedel et al., 2007) and a 'basic pre-requisite' for growth and adoption of new technology (Reffat, 2003). It is considered a key challenge for the implementation of technology (Shannak, 2013). Assessing awareness is in agreement with Hall and Hord (2011) who presented the seven stages of concern in the adoption of new technology and argued that stage-0 is the awareness stage. This implies that when an individual intends to adopt a technology, he/she must be aware of the advantages and disadvantages of the technology (Rehman et al., 2012). In the light of the definition of technology awareness understanding by Rogers (1995), this study builds and uses a new variable '*technology awareness*', and defines it as the instructor's knowledge about the existence, features, benefit, and using the LMS in his teaching (Faruq & Ahmad, 2013). Some researchers (Faruq & Ahmad, 2013; Rehman et al., 2012; Sun, & Fang, 2016) used awareness as a moderator. In this study 'technology awareness' is considered as a *moderator* between independent and dependent variables of technology adoption model (i.e., UTAUT2).

5.8.5 Culture (As a Moderating Variable)

In a study, Baptista and Oliveira (2015) used Hofstede's cultural dimensions as moderator of the technology adoption and found that individualism/collectivism, power distance, and uncertainty avoidance have a strong influence on user's intention to use the technology. Baptista and Oliveira, (2015) found that Venkatesh et al. (2012) combined Hofstede's cultural dimensions as moderators with the UTAUT2 model in the acceptance of mobile banking to achieve the strengths of two theories. Culture is a substantial moderator of technology adoption (Gallivan & Srite, 2005; Im, Hong, & Kang, 2011; Martins, Oliveira, & Popovič, 2014) in the context of mobile banking (Abdullah et al., 2016). Rajapakse (2011) proved that "*culture is a direct construct of behavioral intention*". It was proposed by Rajapakse (2011) to extend the UTAUT model by including culture as the fifth determinant, whereas much other research on the adoption of technology has considered culture as one of the possible moderating variables (Venkatesh & Zhang, 2010) that can accelerate or slow down the adoption process (Robey & Rodriguez, 1989). In addition to the validation of UTAUT, Al-Gahtani et al. (2007) also examined the cultural differences and similarities between Saudi Arabia and North America. Al-Gahtani et al. (2007) and Venkatesh et al. (2003) determined that performance expectancy (PE) has a significant effect on behavioral intention, but they found no moderation of age or gender between intention and behavioral intention. Al-Gahtani et al. (2007) also discovered that effort expectancy (EE) has insignificant impact on behavior intention (BI) in the presence of moderating variables. In UTAUT2, cultural moderators were integrated to evaluate the impact of culture in a mobile banking context (Baptista & Oliveira, 2015). They discovered that power distance, uncertainty avoidance, and collectivism were the most significant cultural moderators of variables of UTAUT2. Saudi culture was added to the list of moderating variables by Alzahrani and Goodwin, (2012). In an African country, Mozambique, Hofstede's

dimensions as moderators were combined with the UTAUT2 model in a mobile banking context. Baptista and Oliveira (2015) supported the integration of more theories to achieve the strengths of different theories. Alzahrani and Goodwin (2012) also supported that cultural construct for KSA can be added as moderating variables. In order to assess the UTAUT model in relation to Hofstede's (1980) cultural dimensions in non-Western countries, AbuShanab, Pearson and Setterstrom (2009) conducted a quantitative study of over 500 Internet banking customers in three Jordanian banks. The results of their study confirmed the cultural influence on technology adoption. Their study indicated that effort expectancy (EE), performance expectancy (PE), and social influence (SI) were moderated by gender. The results also showed that the impact of performance expectancy (PE) is stronger for males, and the impact of social influence (SI) and effort expectancy (EE) is stronger for females, confirming the results of Al-Gahtani et al. (2007). Al-Gahtani et al. (2007) argued that the low individualism index for KSA indicates a strong relationship between behavioral intentions and subjective norms in the Arab world. Although Al-Gahtani et al. (2007) conducted their study on Saudi culture and examined the UTAUT model. Their study was not a longitudinal study of design and needed further validation (Hew et al., 2016). The above review of the literature supports the fact that cultural aspects are missing from most of the technology adoption models, especially from the UTAUT2 model.

5.9 Research Questions

The overall objective of this study was to assess the predictors and moderators of the proposed model that influence the adoption and the use of LMS technology among instructors in the HEIs. Based on the above discussion, the following research questions (RQ) and theoretical framework are derived.

Main Research Question: To what extent do independent variables (such as EE, PE, FC, SI HM, and H) and moderating variables (such as age, experience, gender, technology awareness, and cultural dimensions) of the proposed model influence instructors' behavioral intentions to use an LMS in the HEIs? The main RQ has the following sub-research questions:

RQ 1: To what extent (if any) is behavioral intention (BI) a predictor of use behavior (UB) of LMS technology at HEIs?
RQ 2: To what extent (if any) do independent variables (EE, PE, SI, FC, HM, and H) impact instructors' behavioral intentions to adopt an LMS at HEIs?

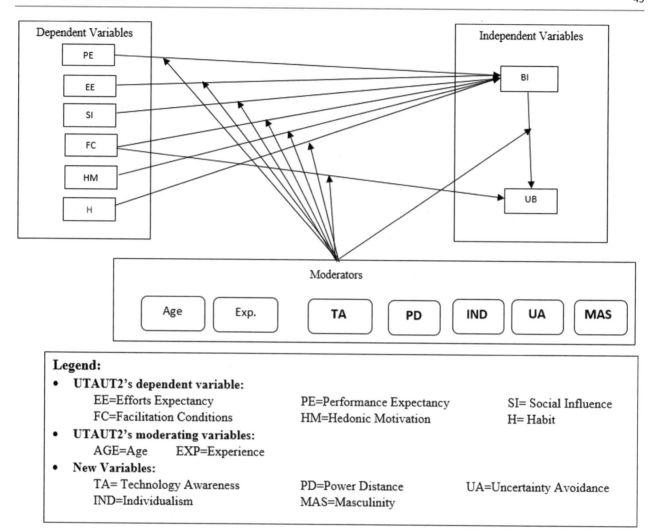

Fig. 5.14 Proposed model—an extension of the UTAUT2 model. Adapted from Venkatesh et al. (2012)

RQ 3: Which out of the six independent variables (PE, EE, FC, SI, HM, and H) delivers the most significant contribution to instructors' behavioral intentions to adopt an LMS at HEIs?

RQ 4: To what extent (if any) do moderating variables, moderate the relationship between the dependent and independent variables?

5.10 Toward a Theoretical Framework

The development of the proposed model is based on the literature on technology adoption models, the UTAUT model by Venkatesh et al. (2003) and the UTAUT2 model

by Venkatesh et al. (2012). The proposed model is shown in Fig. 5.14. The new variables in the proposed model are highlighted as bold type.

5.11 Summary

The study has developed a conceptual framework by adapting the UTAUT2 model by the inclusion of 'Hofstede's cultural dimension' as moderators in the cultural context HEIs. We extended the UTAUT2 model with 'technology awareness' and 'Hofstede's cultural dimensions' as moderating variables of the UTAUT2 model. Finally, this study examined the adequacy of the original UTAUT2 model in higher educational institutions.

References

Abdullah, M., & Khanam, L. (2016). The Influence of Website Quality on m-banking services Adoption in Bangladesh : Applying the UTAUT2 model using PLS. In *International Conference on Electrical, Electronics, and Optimization Techniques (ICEEOT)* (pp. 1–19).

Abdullah, M., Hu, W., & Khanam, L. (2016). The influence of Website quality on m-banking services adoption in Bangladesh : Applying the UTAUT2 model using PLS. In *International Conference on Electrical, Electronics, and Optimization Techniques (ICEEOT)* (pp. 1–19).

Abu-Shanab, E., & Pearson, J. M. (2009). Internet banking in Jordan: An Arabic instrument validation process. *The International Arab Journal of Information Technology, 6*(3), 235–244.

AbuShanab, E., Pearson, J., & Setterstrom, A. J. (2009). Internet banking and customers' acceptance in Jordan: The unified model's perspective. *Communications of AIS, 26,* 493–524.

AbuShanab, E., Pearson, J. M., & Setterstrom, A. (2010). Internet banking and customers' acceptance in Jordan: The unified model's perspective. *Communications of the Association for Information Systems, 26*(1), 493–524.

Agarwal, R., & Prasad, J. (1999). Are individual differences Germane to the acceptance of new information technologies? *Decision Sciences, 30,* 361–391.

Aggelidis, V. P., & Chatzoglou, P. D. (2009). Using a modified technology acceptance model in hospitals. *International Journal of Medical Informatics, 78*(2), 115–126. https://doi.org/10.1016/j.ijmedinf.2008.06.006.

Ahmed, U., Zin, L. M., Halim, A., & Majid, A. (2016). Impact of intention and technology awareness on transport industry's E-service: Evidence from an emerging economy. *International Journal of Industrial Distribution & Business, 7*(3), 13–18.

Ajzen, I. (1991). The theory of planned behavior. *Organizational Behavior and Human Decision Processes, 50*(2), 179–211.

Ajzen, I., & Fishbein, M. (1980). *Understanding attitudes and predicting social behavior.* Englewood Cliffs, NJ: Prentice-Hall.

AlAwadhi, S., & Morris, A. (2008). The use of the UTAUT model in the adoption of E-Government services in Kuwait. In *Proceedings of the 41st Annual Hawaii International Conference on System Sciences (HICSS 2008)* (pp. 219–219). https://doi.org/10.1109/HICSS.2008.452.

Alazzam, M. B., Basari, A. S. H., Sibghatullah, A. S., Ibrahim, Y. M., Ramli, M. R., & Naim, M. H. (2016). Trust in stored data in EHRs acceptance of medical staff: Using UTAUT2. *International Journal of Applied Engineering Research, 11*(4), 2737–2748.

Alba, J. W., & Hutchinson, J. W. (1987). Dimensions of consumer expertise. *Journal of Consumer Research, 13*(March), 411. https://doi.org/10.1086/209080.

Alice, W. M. (2011). *Towards adoption of electronic learning: An empirical investigation of faculty behavioral intentions.* Capella University.

Al-Gahtani, S. S. (2003). Computer technology adoption in Saudi Arabia: Correlates of perceived innovation attributes. *Information Technology for Development, 10*(1), 57–69. https://doi.org/10.1002/itdj.1590100106.

Al-Gahtani, S. S., Hubona, G. S., & Wang, J. (2007). Information Technology (IT) in Saudi Arabia: Culture and the acceptance and use of IT. *Information and Management, 44*(8), 681–691. https://doi.org/10.1016/j.im.2007.09.002.

Alghamdi, S. R. (2016). Use and attitude towards Learning Management Systems (LMS) in Saudi Arabian universities. *Eurasia Journal of Mathematics, Science and Technology Education, 12*(9), 2309–2330. https://doi.org/10.12973/eurasia.2016.1281a.

Al-Qeisi, K. (2009). Analyzing the Use of UTAUT model in explaining an online behavior: Internet Banking Adoption [Brunel University]. http://bura.brunel.ac.uk/bitstream/2438/3620/1/KholoudThesis.pdf.

Al-sobhi, F., Weerakkody, V., & El-Haddadeh, R. (2011). The relative importance of intermediaries in egovernment adoption : A study of Saudi Arabia (pp. 62–74).

Alzahrani, M. E., & Goodwin, R. D. (2012). Towards a UTAUT-based model for the study of E-Government Citizen acceptance in Saudi Arabia (pp. 8–15).

Anderson, C. S., Al-Gahtani, S., & Hubona, G. (2011). The value of TAM antecedents in Global IS development and research. *Journal of Organizational and End User Computing, 23*(1), 18–37. https://doi.org/10.4018/joeuc.2011010102.

Bagozzi, Davis, F. D., & Warshaw, P. R. (1992). Development and test of a theory of technological learning and usage. *Human Relations, 45*(7), 659–686.

Bandura, A. (1986). *Social foundations of thought and action: A social cognitive theory.* Englewood Cliffs, NJ: Prentice Hall.

Bandyopadhyay, K., & Fraccastoro, K. A. (2007). The effect of culture on user acceptance of information technology. *Communications of AIS, 19,* 522–543.

Baptista, G., & Oliveira, T. (2015). Understanding mobile banking: The unified theory of acceptance and use of technology combined with cultural moderators. *Computers in Human Behavior, 50,* 418–430. https://doi.org/10.1016/j.chb.2015.04.024.

Bardram, J. E., & Hansen, T. R. (2010). Context-based workplace awareness. *Computer Supported Cooperative Work (CSCW), 19*(2), 105–138.

Baron, R. M., & Kenny, D. A. (1986). The moderator-mediator variable distinction in social psychological research: Conceptual, strategic, and statistical considerations. *Journal of Personality and Social Psychology, 51*(6), 1173–1182.

Baron, S., Patterson, A., & Harris, K. (2006). Beyond technology acceptance—understanding consumer practice. *International Journal of Service Industry Management, 17*(2), 111–135. https://doi.org/10.1108/09564230610656962.

Berthon, P., Pitt, L., Ewing, M., & Carr, C. (2002). Potential research space in MIS: A framework for envisioning and evaluating research replication, extension, and generation. *Information Systems Research, 13*(4), 416.

Brown, S. A., & Venkatesh, V. (2005). Model of adoption of technology in households: A baseline model test and extension incorporating household life cycle. *MIS Quarterly, 29*(4), 399–426.

Burton-Jones, A., & Grange, C. (2013). From use to effective use: A representation theory perspective. *Information Systems Research, 24*(3), 632–658.

Chang, I. C., Hwang, H. G., Hung, W. F., & Li, Y. C. (2007). Physicians' acceptance of pharmacokinetics-based clinical decision support systems. *Expert Systems with Applications, 33*(2), 296–303.

Charbaji, A., & Mikdashi, T. (2003). A path analytic study of the attitude towards e-government in Lebanon. *Corporate Governance, 3*(1), 76–82.

Chau, P., & Hu, P. (2002). Examining a model of information technology acceptance by individual professionals: An exploratory study. *Journal of Management Information Systems, 18*(4), 191–229.

Chen, C. P., Lai, H. M., & Ho, C. Y. (2015). Why do teachers continue to use teaching blogs? The roles of perceived voluntariness and habit. *Computers and Education, 82*(1), 236–249. https://doi.org/10.1016/j.compedu.2014.11.017.

Cheng, B., Wang, M., Moormann, J., Olaniran, B. A., & Chen, N.-S. (2012). The effects of organizational learning environment factors on e-learning acceptance. *Computers and Education, 58*(3), 885–899. https://doi.org/10.1016/j.compedu.2011.10.014.

Chiu, C. Y., Leung, A. K. Y., & Hong, Y. Y. (2010). *Cultural processes: An overview*. Cambridge University Press.

Chiu, C. M., & Wang, E. T. (2008). Understanding web-based learning continuance intention: The role of subjective task value. *Information & Management, 45*(3), 194–201.

Compeau, D. R., & Higgins, C. A. (1995). Computer self-efficacy: Development of a measure and intitial test. *MIS Quarterly, 19,* 189–211.

Curtis, L., Edwards, C., Fraser, K. L., Gudelsky, S., Holmquist, J., Thornton, K., & Sweetser, K. D. (2010). Adoption of social media for public relations by nonprofit organizations. *Public Relations Review, 36*(1), 90–92.

Dadayan, L., & Ferro, E. (2005). When technology meets the mind: A comparative study of the technology acceptance model. In *International Conference on Electronic Government* (pp. 137–144).

Davis, F. D. (1989). Perceived usefulness, perceived ease of use, and user acceptance of information technology. *MIS Quarterly*, 319–340.

Davis, F. D., Bagozzi P. R., & Warshaw, P. R. (1989). User acceptance of computer technology: a comparison of two theoretical models. *Management Science, 35*(8), 982–1003.

Davis, F. D., & Venkatesh, V. (1996). A critical assessment of potential measurement biases in the technology acceptance model: Three experiments. *International Journal of Human-Computer Studies, 45*(1), 19–45. https://doi.org/10.1006/ijhc.1996.0040.

Dulle, F. W., & Minishi-Majanja, M. K. (2011). The suitability of the unified theory of acceptance and use of technology (UTAUT) model in open access adoption studies.*Information Development, 27*(1), 32–45. https://doi.org/10.1177/0266666910385375.

De Wit, K., Heerwegh, D., & Verhoeven, J. C. (2014). Can openness to ICT and scientific research predict the ICT skills and ICT use of bachelor's students? *Computers and Education, 78,* 397–413. https://doi.org/10.1016/j.compedu.2014.07.003.

Deci, E. L. (1975). *Intrinsic motivation*. Plenum Press.

Deci, E. L., & Ryan, R. M. (1987). The support of autonomy and the control of behavior. *Journal of Personality and Social Psychology, 53*(6), 1024–1037.

Diep, N. A., Cocquyt, C., Zhu, C., & Vanwing, T. (2016). Predicting adult learners' online participation: Effects of altruism, performance expectancy, and social capital. *Computers and Education, 101,* 84–101. https://doi.org/10.1016/j.compedu.2016.06.002.

Dishaw, M. T., & Strong, D. M. (1999). Extending the technology acceptance model with task-technology fit constructs. *Information and Management, 36*(1), 9–21.

Dwivedi, Y. K., Rana, N. P., Chen, H., & Williams, M. D. (2011). A Meta-analysis of the Unified Theory of Acceptance and Use of Technology (UTAUT). 155–170.

Eckhardt, A., Laumer, S., & Weitzel, T. (2009). Who influences whom? Analyzing workplace referents' social influence on IT adoption and non-adoption. *Journal of Information Technology, 24*(1), 11–24.

Ellis, E. R., & Allaire, A. J. (1999). Modeling computer interest in older adults: the role of age, education, computer knowledge, and computer anxiety. *Human Factors, 41,* 345–355.

Faruq, M. A., & Ahmad, H. B. (2013). The moderating effect of technology awareness on the relationship between UTAUT constructs and behavioural intention to use technology: A conceptual paper. *Australian Journal of Business and Management Research, 3*(02), 14–23. https://doi.org/10.1016/j.im.2013.09.002.

Fishbein, M., & Ajzen, I. (1975). *Belief, attitude, intention, and behavior: An introduction to theory and research*. Don Mills, Ontario: Addison-Wesley Publishing Company.

Gallivan, M., & Srite, M. (2005). Information technology and culture: Identifying fragmentary and holistic perspectives of culture. *Information and Organization, 15*(2), 295–338. https://doi.org/10.1016/j.infoandorg.2005.02.005.

Ghobakhloo, M., Zulkifli, N. B., & Aziz, F. A. (2010). The interactive model of user information technology acceptance and satisfaction in small and medium-sized enterprises. *European Journal of Economics, Finance and Administrative Sciences, 19.*

Gholami, R., Ogun, A., Koh, E., & Lim, J. (2010). Factors affecting e-payment adoption in Nigeria. *Journal of Electronic Commerce in Organizations (JECO), 8*(4), 51–67.

Giannakos, M. N., & Vlamos, P. (2013). Educational webcasts' acceptance: Empirical examination and the role of experience. *British Journal of Educational Technology, 44*(1), 125–143. https://doi.org/10.1111/j.1467-8535.2011.01279.x.

Gilbert, D., Balestrini, P., & Littleboy, D. (2004). Barriers and benefits in the adoption of e-government. *International Journal of Public Sector Management, 17*(4), 286–301. https://doi.org/10.1108/09513550410539794.

Godin, G., & Kok, G. (1996). The theory of planned behaviour: A review of its applications to health-related behaviours. *American Journal of Health Promotion, 11*(2), 87–98.

Gupta, B., Dasgupta, S., & Gupta, A. (2008). Adoption of ICT in a government organization in a developing country: An empirical study. *The Journal of Strategic Information Systems, 17*(2), 140–154. https://doi.org/10.1016/j.jsis.2007.12.004.

Hadji, B., & Degoulet, P. (2016). Information system end-user satisfaction and continuance intention: A unified modeling approach. *Journal of Biomedical Informatics, 61,* 185–193. https://doi.org/10.1016/j.jbi.2016.03.021.

Hall, G. E., & Hord, S. M. (2011). *Implementing change: Patterns, principles, and potholes* (3rd ed.). Pearson Education.

Hall, D., & Mansfield, R. (1975). Relationships of age and seniority with career variables of engineers and scientists. *Journal of Applied Psychology, 60*(3), 201–210.

Handy, J. H., Whiddett, R., & Hunter, I. (2001). A technology acceptance model for inter-organisational electronic medical records systems. *AJIS, 9*(1).

He, D., & Lu, Y. (2007). Consumers perceptions and acceptances towards mobile advertising: An empirical study in China. In *International Conference on Wireless Communications, Networking and Mobile Computing* (pp. 3775–3778).

Hew, T., Latifah, S., & Abdul, S. (2016). Computers and education understanding cloud-based VLE from the SDT and CET perspectives: Development and validation of a measurement instrument. *Computers and Education, 101,* 132–149. https://doi.org/10.1016/j.compedu.2016.06.004.

Hofstede, G. (1980). *Culture's consequences: International differences in work-related values*. Sage Publications.

Hsu, H. (2012). The acceptance of moodle : An empirical study based on UTAUT. *3*(December), 44–46. https://doi.org/10.4236/ce.2012.38b010.

Hu, P. J., Chau, P. Y., Sheng, O. R. L., & Tam, K. Y. (1999). Examining the technology acceptance model using physician acceptance of telemedicine technology. *Journal of Management Information Systems*, 91–112.

Igbaria, M., Parasuraman, S., & Baroudi, J. J. (1996). A motivational model of microcomputer usage. *Journal of Management Information Systems, 13*(1), 127–143.

Im, Il., Hong, S., & Kang, M. S. (2011). An international comparison of technology adoption testing the UTAUT model. *Information & Management, 48*(1), 1–8. https://doi.org/10.1016/j.im.2010.09.001.

Islam, A. K. M. N. (2011). Understanding the continued usage intention of educators toward an e-learning system. *International Journal of E-Adoption, 3*(2), 54–69. https://doi.org/10.4018/jea.2011040106.

Ismail, A. (2016). The effective adoption of ICT-enabled services in educational institutions–key issues and policy implications. *Journal of Research in Business, Economics and Management (JRBEM), 5*(5), 717–728.

Jayasingh, S., & Eze, U. (2010). The role of moderating factors in mobile coupon adoption: An extended TAM perspective. *Communications of the IBIMA*.

Jen, W., Lu, T., & Liu, P. T. (2009). An integrated analysis of technology acceptance behaviour models: Comparison of three major models. *MIS Review, 15*(1), 89–121.

Jong, D., & Wang, T. S. (2009). Student acceptance of web-based learning system. In *The 2009 International Symposium on Web Information Systems and Applications (WISA 2009)* (p. 53).

Kassarjian, H. H., & Robertson, T. S. (1991). *Perspectives in consumer behaviour* (4th ed.). New Jersey: Prentice Hall.

Keramati, A., Sharif, H. J., Azad, N., & Soofifard, R. (2012). Role of subjective norms and perceived behavioral control of tax payers in acceptance of E-Tax payment system. *International Journal of E-Adoption, 4*(3), 1–14. https://doi.org/10.4018/jea.2012070101.

Kijsanayotin, B., Pannarunothai, S., & Speedie, S. M. (2009). Factors influencing health information technology adoption in Thailand's community health centers: Applying the UTAUT model. *International Journal of Medical Informatics, 78*(6), 404–416. https://doi.org/10.1016/j.ijmedinf.2008.12.005.

Kim, S. S., & Malhotra, N. K. (2005). A longitudinal model of continued is use: An integrative view of four mechanisms underlying postadoption phenomena. *Management Science, 51*(5), 741–755. https://doi.org/10.1287/mnsc.1040.0326.

Kripanont, N. (2007). Using technology acceptance model of Internet usage by academics within Thai business schools [Victoria University]. http://wallaby.vu.edu.au/adt-VVUT/public/adtVUT20070911.152902/index.html.

Lakhal, S., Khechine, H., & Pascot, D. (2013). Student behavioural intentions to use desktop video conferencing in a distance course: integration of autonomy to the UTAUT model. *Journal of Computing in Higher Education, 25*(2), 93–121. https://doi.org/10.1007/s12528-013-9069-3.

Laumer, S., Eckhardt, A., & Weitzel, T. (2010). Electronic human resources management in an e-business environment. *Journal of Electronic Commerce Research, 11*(4), 240.

Lee, J. (2004). Discriminant analysis of technology adoption behavior: A case of internet technologies in small businesses. *Journal of Computer Information Systems, 44*(4), 57–66.

Lee, F.-H., & Wu, W.-Y. (2011). Moderating effects of technology acceptance perspectives on e-service quality formation: Evidence from airline websites in Taiwan. *Expert Systems with Applications, 38*(6), 7766–7773.

Legris, P., Ingham, J., & Collerette, P. (2003). Why do people use information technology? A critical review of the technology acceptance model. *Information and Management, 40*(3), 191–204. https://doi.org/10.1016/S0378-7206(01)00143-4.

Lent, R. W., Brown, S. D., & Hackett, G. (1994). Toward a unifying social cognitive theory of career and academic interest, choice, and performance. *Journal of Vocational Behaviour, 45,* 79–122.

Lim. (2003). A conceptual framework on the adoption of negotiation support systems. *Information and Software Technology, 45,* 469–477.

Limayem, M., Hirt, S. G., & Cheung, C. M. K. (2007). How habit limits the predictive power of intention: The case of information systems continuance. *MIS Quarterly: Management Information Systems, 31*(4), 705–737. https://doi.org/10.2307/25148817.

Lin, C. P., & Anol, B. (2008). Learning online social support: An investigation of network information technology based on UTAUT. *CyberPsychology & Behavior, 11*(3).

Loo, W. H., Yeow, P. H., & Chong, S. C. (2009). User acceptance of Malaysian government multipurpose smartcard applications. *Government Information Quarterly, 26*(2), 358–367.

Luo, M. M., Chea, S., & Chen, J.-S. (2011). Web-based information service adoption: A comparison of the motivational model and the uses and gratifications theory. *Decision Support Systems, 51*(1), 21–30. https://doi.org/10.1016/j.dss.2010.11.015.

Marques, B. P., Villate, J. E., & Carvalho, C. V. (2011). Applying the UTAUT model in engineering higher education: Teacher' s technology adoption. In *6th Iberian conference on information systems and technologies CISTI 2011* (pp. 1–6).

Martins, C., Oliveira, T., & Popovič, A. (2014). Understanding the Internet banking adoption: A unified theory of acceptance and use of technology and perceived risk application. *International Journal of Information Management, 34*(1), 1–13.

Mathieson, K., Peacock, E., & Chin, W. W. (2001). Extending the technology acceptance model: The influence of perceived user resources. *Database for Advances in Information Systems, 32*(3), 86.

Mohammadyari, S., & Singh, H. (2015). Computers and Education Understanding the effect of e-learning on individual performance: The role of digital literacy. *Computers and Education, 82,* 11–25. https://doi.org/10.1016/j.compedu.2014.10.025.

Moore, G. C., & Benbasat, I. (1991). Develoment of an instrument to measure the perceptions of adopting an information technology innovation. *Information Systems Research, 2*(3).

Morris, M. G., & Venkatesh, V. (2000). Age differences in technology adoption decisions: Implications for a changing work force. *Personnel Psychology, 53*(2), 375–403. https://doi.org/10.1111/j.1744-6570.2000.tb00206.x.

N Bourbon, I., & Hollet-Haudebert, S. (2009). Pourquoi contribuer à des bases de connaissances? Une exploration des facteurs explicatifs à la lumière du modèle UTAUT. *Systèmes d'Information et Management, 14*(1), 9–36.

Negahban, A., & Chung, C. -H. (2014). Discovering determinants of users perception of mobile device functionality fit. *Computers in Human Behavior, 35,* 75–84. https://doi.org/10.1016/j.chb.2014.02.020.

Nov, O., & Ye, C. (2009). Resistance to change and the adoption of digital libraries: An integrative model. *Journal of the American Society for Information Science and Technology, 60*(8), 1702–1708.

Novakovic, L., McGill, T., & Dixon, M. (2009). Understanding user behavior towards passwords through acceptance and use modelling. *International Journal of Information Security and Privacy, 3*(1), 11–29. https://doi.org/10.4018/jisp.2009010102.

Nysveen, H., Pedersen, P. E., & Thorbjørnsen, H. (2005). Intentions to use mobile services: Anteced-ents and cross- service comparisons. *Journal of the Academy of Marketing Science, 33*(3), 330–346. https://doi.org/10.1177/0092070305276149.

Oliveira, T., & Martins, M. F. (2011). Literature review of information technology adoption models at firm level. *The Electronic Journal Information Systems Evaluation (EJISE), 14*(1), 110–121.

Oyelaran-Oyeyinka, B., & Adeya, C. N. (2003). Dynamics of adoption and usage of ICTs in African universities: A study of Kenya and Nigeria. *Technovation, 24*(10), 841–851.

Pajares, F., & Schunk, D. H. (2002). Self and self-belief in psychology and education: A historical perspective. Improving academic achievement. In *Impact of psychological factors on education* (pp. 3–21). http://www.uky.edu/~eushe2/Pajares/efftalk.html.

Pearlson, K., & Sauders, C. (2003). *No managing and using information systems: A strategic approach*. New York: Wiley.

Porter, W. W., & Graham, C. R. (2016). Institutional drivers and barriers to faculty adoption of blended learning in higher education. *British Journal of Educational Technology, 47*(4), 748–762. https://doi.org/10.1111/bjet.12269.

Rahim, N. Z. A., Lallmahomed, M. Z., Ibrahim, R., & Rahman, A. A. (2011). No title a preliminary classification of usage measures in information system acceptance: A Q-sort approach. *International Journal of Technology Diffusion, 2*(4), 4–25.

Rajapakse, J. (2011). Extending the Unified Theory of Acceptance and Use of Technology (UTAUT) model. In *2011 4th International Conference on Interaction Sciences (ICIS)* (pp. 47–52).

Reffat, R. (2003). *Developing a successful E-Government (Working Paper)*.

Rehman, M., Esichaikul, V., & Kamal, M. (2012). Factors influencing e-government adoption in Pakistan. *Transforming Government: People, Process and Policy, 6*(3), 258–282.

Riedel, J., Pawar, K., Torroni, S., & Ferrari, E. (2007). A survey of RFID awareness and use in UK logistics industry. In *Dynamics in logistics, first international conference, LDIC 2007* (pp. 105–115).

Robey, E., & Rodriguez-Diaz, A. (1989). The organizational and cultural context of systems implementation: case experience for Latin America. *Information Management, 17*(4), 229–239.

Rodrigues, G., Sarabdeen, J., & Balasubramanian, S. (2016). Factors that influence consumer adoption of e-government services in the UAE: A UTAUT model perspective. *Journal of Internet Commerce, 15*(1), 18–39. https://doi.org/10.1080/15332861.2015.1121460.

Rogers, E. M. (1983). *Diffusion of innovations*. Free Press.

Rogers, E. M. (1995). *Diffusion of innovation* (5th ed.). The Free Press.

Rogers, E. M. (2003). *Diffusion of innovations* (5th ed). Free Press.

San Martin, H., & Herrero, A. (2012). Influence of the user's psychological factors on the online purchase intention in rural tourism: Integrating innovativeness to the UTAUT framework. *Tourism Management, 33*(2), 341–350.

Schaper, L. K., & Pervan, G. P. (2007). ICT and OTs: A model of information and communication technology acceptance and utilisation by occupational therapists. *International Journal of Medical Informatics, 76*, S212–S221.

Schaupp, L. C., Carter, L., & McBride, M. E. (2010). E-file adoption: A study of US taxpayers' intentions. *Computers in Human Behavior, 26*(4), 636–644.

Schifter, D. E., & Ajzen, I. (1985). Intention, perceived control, and weight loss: An application of the theory of planned behaviour. *Journal of Personality and Social Psychology, 49*, 843–851.

Serenko, A., Turel, O. & Yol, S. (2006). Moderating roles of user demographics in the American customer satisfaction model within the context of mobile services. *Journal of Information Technology Management, 17*(4). http://jitm.ubalt.edu/xvii-4/article3.pdf.

Shannak, R. O. (2013). The difficulties and possibilities of E-Government: The case of Jordan. *Journal of Management Research, 5*(2), 189–204. https://doi.org/10.5296/jmr.v5i2.2560.

Shareef, M. A., Kumar, V., Kumar, U., & Dwivedi, Y. K. (2011). e-Government Adoption Model (GAM): Differing service maturity levels. *Government Information Quarterly, 28*(1), 17–35. https://doi.org/10.1016/j.giq.2010.05.006.

Sheppard, B. H., Hartwick, J., & Warshaw, P. R. (1988). The theory of reasoned action: A meta-analysis of past research with recommendations for modifications and future research. *The Journal of Consumer Research, 15*(3), 325–343.

Shiau, W. L., & Chau, P. Y. K. (2015). Understanding behavioral intention to use a cloud computing classroom: A multiple model comparison approach. *Information and Management, 53*(3), 355–365. https://doi.org/10.1016/j.im.2015.10.004.

Shin, D. H. (2009). Towards an understanding of the consumer acceptance of mobile wallet. *Computers in Human Behavior, 25*(6), 1343–1354.

Slade, E. L. & Williams, M. (2013). An extension of the UTAUT 2 in a healthcare context. In *Proceeding of the UK Academy for Information Systems*.

Straub, E. T. (2009). Understanding technology adoption: Theory and future directions for informal learning. *Review of Educational Research, 79*(2), 625–649.

Sun, H., & Fang, Y. (2016). Choosing a fit technology: Understanding mindfulness in technology adoption and continuance. *Journal of the Association for Information Systems, 17*(6), 377.

Sun, Y., Bhattacherjee, A., & Ma, Q. (2009). Extending technology usage to work settings: The role of perceived work compatibility in ERP implementation. *Information and Management, 46*(6), 351–356. https://doi.org/10.1016/j.im.2009.06.003.

Taylor, S., & Todd, P. (1995). Understanding information technology usage: A test of competing models. *Information Systems Research, 6*(4), 144–176.

Thompson, R. L., Higgins, H. (1991). Personal computing: Toward a conceptual model of utilization. *MIS Quarterly, 15*(1), 125–142.

Thong, Y. L. J., Hong, S. -J., & Tam, K. Y. (2006). The effects of post-adoption beliefs on the expectation-confirmation model for information technology continuance. *International Journal of Human-Computer Studies, 64*(9), 799–810.

Tibenderana, P. K., & Ogao, P. (2009). Information technologies acceptance and use among universities in Uganda: A model for hybrid library services end-users. *International Journal of Computing and ICT Research, 1*(1), 60–75. www.ijcir.org/.

Triandis, H. C. (1977). *Interpersonal behavior*. Monterey, CA: Brooks/Cole Publishing Company.

Van Biljon, J. A. (2006). *A model for representing the motivaitonal and cultural factors that influence mobile phone useage variety*. Unpublished Dissertation, University of South Africa.

Van Biljon, J., & Kotze, P. (2007). Modeling the factors that influence mobile phone adoption. *South African Institute of Computer Scientists and Information Technologists*, 2–3.

Van Biljon, J., & Kotzé, P. (2008). Cultural factors in a mobile phone adoption and usage model. *14*(16), 2650–2679.

Van der Heijden, H. (2004). User acceptance of hedonic information systems. *MIS Quarterly*, 695–704.

Van Dijk, J. A., Peters, O., & Ebbers, W. (2008). Explaining the acceptance and use of government Internet services: A multivariate analysis of 2006 survey data in the Netherlands. *Government Information Quarterly, 25*(3), 379–399.

Vatanparast, R. (2010). Theories behind mobile marketing research (pp. 255–278). https://doi.org/10.4018/978-1-60566-074-5.ch014.

Venkatesh, V., & Bala, H. (2008). Technology acceptance model 3 and a research agenda on interventions. *Decision Sciences, 39*(2), 273–315. https://doi.org/10.1111/j.1540-5915.2008.00192.x.

Venkatesh, V., & Davis, F. D. (2000). A theoretical extension of the technology acceptance model: Four longitudinal field studies. *Management Science, 46*(2), 186–204.

Venkatesh, V., Davis, F., & Morris, M. G. (2007). Dead or alive? The development, trajectory and future of technology adoption research. *Journal of the Association for Information Systems, 8*(4).

Venkatesh, V., & Speier, C. (1999). Computer technology training in the workplace: A longitudinal investigation of the effect of mood. *Organizational Behavior and Human Decision Processes, 79*, 1–28.

Venkatesh, Viswanath, Speier, C., & Morris, M. G. (2002). User acceptance enablers in individual decision making about technology: Toward an integrated model. *Decision Sciences, 33*(2), 297–316. https://doi.org/10.1111/j.1540-5915.2002.tb01646.x.

Venkatesh, V., Morris, M. G., Davis, G. B., & Davis, F. D. (2003). User acceptance of information technology: Toward a unified view. *27*(3), 425–478.

Venkatesh, V., Thong, J., & Xu, X. (2012). Consumer acceptance and user of information technology: Extending the unified theory of acceptance and use of technology. *MIS Quarterly, 36*(1), 157–178. http://ezproxy.library.capella.edu/login;   http://search.ebscohost.

com/login.aspx?direct=true&db=iih&AN=71154941&site=ehost-live&scope=site.

Venkatesh, V., & Zhang, Z. (2010). Unified theory of acceptance and use of technology: U.S. vs. China. *Journal of Global Information Technology Management, 13*(1), 5–12.

Vroom, V. H. (1964). *Work and motivation* (p. 54). New York: Wiley.

Wakefield, R., & Whitten, D. (2006). Mobile computing: a user study on hedonic/utilitarian mobile device usage. *European Journal of Information Systems, 5*(1), 292–300. https://doi.org/10.1057/palgrave.ejis.3000619.

Williams, P. W. (2009). *Assessing mobile learning effectiveness and acceptance dissertation directed by.*

Wong, K.-T., Teo, T., & Russo, S. (2012). Interactive whiteboard acceptance: Applicability of the UTAUT model to student teachers. *The Asia-Pacific Education Researcher, 22*(1), 1–10. https://doi.org/10.1007/s40299-012-0001-9.

Wu, H., Hsu, Y., & Hwang, F. (2007). *Factors affecting teachers' adoption of technology. September 2006* (pp. 63–85).

Wu, Y., Tao, Y., & Yang, P. (2007). Using UTAUT to explore the behavior of 3G mobile communication users. In *2007 IEEE international conference on industrial engineering and engineering management* (pp. 199–203). https://doi.org/10.1109/IEEM.2007.4419179.

Ye, C., Seo, D., Desouza, K. C., Sangareddy, S. P., & Jha, S. (2008). Influences of IT substitutes and user experience on post-adoption user switching: An empirical investigation. *Journal of the American Society for Information Science and Technology, 59*(13), 2115–2132.

YenYuen, Y., & Yeow, P. H. P. (2009). User acceptance of internet banking service in Malaysia. *Web Information Systems and Technologies, 18*, 295–306.

Yeow, P. H. P., & Loo, W. H. (2009). Acceptability of ATM and transit applications embedded in multipurpose smart identity Card : An exploratory study in malaysia. *5.*

Yeow, P. H. P., Yuen, Y. Y., Tong, D. Y. K., & Lim, N. (2008). User acceptance of online banking service in Australia. *International Business Information Management Assoc-Ibima, IBIM Norristown, 1*(22), 191–197.

Yi, M. Y., Jackson, J. D., Park, J. S., & Probst, J. C. (2006). Understanding information technology acceptance by individual professionals: Toward an integrative view. *Information and Management, 43*(3), 350–363. https://doi.org/10.1016/j.im.2005.08.006.

Zanjani, N., Edwards, S. L., Nykvist, S., & Geva, S. (2016). LMS acceptance: The instructor role. *Asia-Pacific Education Researcher, 25*(4), 519–526. https://doi.org/10.1007/s40299-016-0277-2.

Zhou, T., Lu, Y., & Wang, B. (2010). Integrating TTF and UTAUT to explain mobile banking user adoption. *Computers in Human Behavior, 26*(4), 760–767. https://doi.org/10.1016/j.chb.2010.01.013.

Zolfaghar, K., Khoshalhan, F., & Rabiei, M. (2010). User acceptance of location-based mobile advertising. *International Journal of E-Adoption, 2*(2), 35–47. https://doi.org/10.4018/jea.2010040103.

6.1 Introduction

This chapter explains the chosen methodology by emphasizing the important features, strengths, and weaknesses of various approaches and methods. It includes the research design, survey instruments, data collection, and statistical techniques for analyses. It also describes the procedure followed and the motives behind the specific techniques, methods, and approaches of data collection. Finally, this chapter discusses the reliability, validity, and associated issues for the instruments used for data collection and analysis.

6.2 What Is the Research Methodology?

In general, any research study is driven by the research problem. The nature of the research problem generally leads to the data required, how the data should be collected, where the data should be collected, and the kind of statistical analyses that will be carried out. Various definitions have been used in the literature to describe the meanings of the research process or methodology. Saunders, Lewis, and Thornhill (2009) described that research is a process of data collection in an organized way and then interpreting the results to answer the research problem and to achieve the objective of the research. Thus the main objective of any research study is to answer specific research questions (Yates, 2004). Robson (2002) classified research types into descriptive, exploratory, and explanatory. The purpose of the descriptive study is "to portray an accurate profile of persons, events or situations" (Robson, 2002, p. 59). Descriptive research is conducted to 'describe' the detailed measurement and report on the characteristics of a phenomenon or population under research. Such studies are usually conducted by surveys. The purpose of the exploratory study is to understand the issue, especially when the reasons for the problem are unknown due to a lack of research design or lack of information about the subject matter or lack of existing research (Saunders et al., 2009). Good exploratory research generally starts with a broader view of a problem, then it narrows down gradually as the study proceeds. In an explanatory research design, the relationships between variables are identified and analyzed to understand how and why some phenomena happened. In explanatory research, the researcher can control the variables to examine the effect of one or more variables on the problem (Hussey & Hussey, 1997). This current study examines the relationship between the independent and the dependent variables. Hence, it is an explanatory research study.

6.3 Research Design

A research design is a strategy or a complete logic of the research to obtain the answers to the research question (Yin, 2003, p. 19). The research design is "a basic set of beliefs that guide action" (Guba, 1990, p. 17). Saunders et al. described that the research process is a series of stages through which one must pass to answer his or her research question and to complete the research project. In general, these stages include formulating a research topic, literature review, designing the research, data collection, data analyzing, and finally writing (2016, p. 11). The following research onion (Fig. 6.1) is adapted from a book by Saunders, Lewis and Thornhill (2016), titled 'Research Methods for Business Students' as a guide to developing a good research methodology.

Saunders et al. (2016) describe the entire research procedure as an 'onion' having six layers. Each layer needs to be peeled off to get to the principal layer, which is the data collection and analysis layer. These layers are 1–philosophy, 2–approach to theory development, 3–methodological, 4–strategy, 5–time horizons, and finally, 6–techniques to collect and then analyze the data.

Thus, the layered 'onion' of Saunders et al. (2016) is used as a guide to explain the methodology used in this study.

© Springer Nature Switzerland AG 2021
R. A. Khan and H. Qudrat-Ullah, *Adoption of LMS in Higher Educational Institutions of the Middle East*,
Advances in Science, Technology & Innovation, https://doi.org/10.1007/978-3-030-50112-9_6

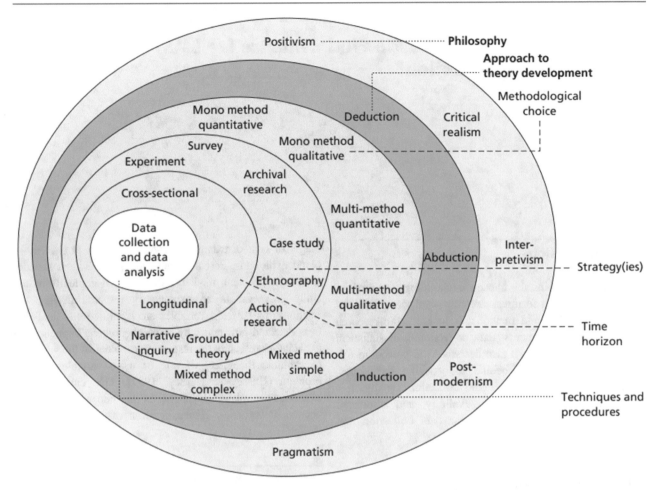

Fig. 6.1 The research onion. *Source* Saunders et al. (2016, p. 124)

The aim of providing information on the various research approaches and philosophies is to elicit the strengths and weaknesses of each, which leads the choice of a specific paradigm for this research.

6.3.1   First Layer: Philosophy

The first basic factor of the research design is a philosophical worldview in terms of the development of knowledge that any researcher considers for the research study (Saunders et al., 2009, 2016). There is no 'one' agreed philosophy. 'Pluralism' and 'Unificationism' were two important perspectives of the debate. The *unificationists* support uniting the research problem under a strong research paradigm, philosophy, and methodology (Saunders et al., 2016, p. 126). Whereas *pluralists* support the diversity and believe that diversity of the fields will improve business and management (Knudsen, 2002). In the research context, it is also important to understand the ontology, epistemology, and axiology. *Ontology* deals with the nature of realities we

encounter in our research. *Epistemology* deals with what constitutes acceptable and valid knowledge that can be communicated to another human. However, "the extent and ways your values influence your research process" is the *axiological* assumption, which includes the role of ethics and values in the research process (Saunders et al., 2016, p. 128). In natural sciences, the '*objectivism*' theorizes that social reality is external to the actors–that is, to the researcher and others. In other words, ontologically, objectivism holds realism; the experiences and interpretations of researchers (social actors) do not influence the social world. Therefore, an objectivist considers that there is one true reality seen throughout an entire society, phenomena exist independently, and the reality is unchanged (Burell & Morgan, 1979). From an epistemological point of view, objectivists try to find the truth about the social world, through measurable facts, which helps to draw law-like generalizations about the world. From the axiological point of view, the social actors (researchers) exist independent of each other; objectivists try to keep their research free of their own beliefs and values.

In the first layer, the following philosophies have been defined by Saunders et al. (2016): positivism, critical realism, interpretivism, postmodernism, and pragmatism. *Positivism philosophy* is related to the concept of objectivity or objectivism. It mainly deals with the quantitative approach with a structured data collection on a large data sample. In this philosophy, the beliefs of a researcher do not impact the study (Saunders et al., 2009; 2016). The positivism philosophy states that reality survives independently and objectively, whereas interpretivism relies on the construction and reconstruction of social and human interactions. The positivist paradigm is characterized by propositions, hypotheses, models, quantifiable factors, and conclusions derived from the samples of the population (Chen & Hirschheim, 2004). The positivist approach is associated with quantitative research and uses numeric data to make generalizations about responses, which are grounded in an epistemology of objectivism (Creswell, 2009). Realism philosophy deals with the reality that exists. It is independent of the mind. Realism has two parts: critical realism and direct realism (Saunders et al., 2009). In direct realism, it is believed that what the researchers see is true. In critical realism, the researchers' experience is considered a sensation (Sekaran & Bougie, 2010). The purpose of the *interpretive* research is to create new, richer understandings and interpretations of social worlds and contexts. Concerning philosophical debates, the constructivist approach is associated with qualitative research which is consistent with an interpretive approach (QiYing Su, Adams, 2010). From an axiological view of interpretivism is that interpretivism understands that the researcher's own beliefs and interpretations play a key role in the process. *Postmodernists* provide alternative worldviews that have been silenced by dominating perspectives. The dominant means of decisions are not necessarily the 'best' but it is just the voices of a particular group(s) of people at a particular point in time. There is also another side of the perspective that is suppressed and might have the power and potential to create alternative truths. Postmodernists aim to challenge the recognized means of thinking and knowing and to support the suppressed and marginalized ways of knowing that have been previously ignored (Saunders et al., 2016, p. 142). *Pragmatic research* is not new to social or management sciences (Parvaiz, Mufti & Wahab, 2016). It was established in the USA in the early twentieth century by Charles Pierce, William James, and John Dewey (Saunders et al., 2016). It is derived from the Greek literature 'Pragma' which means 'action', from which the words 'practical' and 'practice' originated (Pansiri, 2005; Parvaiz et al., 2016). In English, 'pragmatic' has the meaning of searching for workable solutions to complex human problems (Fishman, 1991). The concern for a pragmatist is to find out 'what works' and what enables solutions to problems (Creswell, 2003). The problem or a research question is the 'central' focus (Creswell, 2003) of this approach.

6.3.2 Second Layer: Research Approach

Saunders et al. (2009, 2016) described three approaches within this layer: the deductive approach, inductive approach, and an abductive approach. The *deductive approach* focuses on quantitative methods. In the deductive approach, a researcher begins with existing theory, forms a hypothesis, collects the data, analyzes the data, and then either rejects or accepts the hypothesis. The deductive approach is founded on scientific rules and is an extremely structured approach. The researcher is independent and selects an adequate sample size to generalize the results of the study (Saunders et al., 2009). The *inductive approach* is primarily related to qualitative methods. This approach begins with observing the phenomena, analyzing the themes, developing the relationship(s), and then at the end formulating a theory (Cavana, Delahaye & Sekaran, 2001). In this research approach, the investigator is part of the study, has a profound understanding of the research process, and interacts with participants to collect qualitative data. In an *inductive approach*, the researcher is more flexible to change research focus as the research proceeds. In the abduction approach, the available data is used to discover a phenomenon, to identify patterns and themes, to modify an existing theory or to develop a new theory which is then tested, mostly through a further collection of data. According to Saunders et al., (2009), Hussey and Hussey (1997), a researcher is free to switch from one approach to another to answer the research question.

6.3.3 Third-Layer: Methodological Choice

Saunders et al. (2016) classified this layer into a mono-method qualitative, mono-method quantitative, multi-method qualitative, multi-method quantitative mixed-method complex, and mixed-method simple. However, Creswell (2009) classified the research strategy into qualitative, quantitative and mixed-methods. Quantitative methods are embedded in the formal and fixed design, where a fixed strategy is utilized for its ability to describe the processes and patterns that they can be connected to the social or organizational structure using statistical analysis (Robson, 2002). Survey questionnaires are used to collect the data in quantitative methods (Chen & Hirschheim, 2004). Johnson and Onwuegbuzie (2004) describe that quantitative research methods are quicker and more accurate.

The results of the quantitative study are independent of the biases of a researcher and have greater credibility. This method concentrates on testing theories or measuring phenomena (Hussey & Hussey, 1997). The quantitative research design can be a single technique for data collection such as a questionnaire and its corresponding analysis. Multi-method uses more than one method (quantitative or qualitative) but they are not mixed. The term 'qualitative research' includes the various processes of inquiry that provide an understanding of how all the parts of procedures work together (Merriam, 1998). This research generally seeks to understand human and social behaviour and is subjective in nature (Collis & Hussey, 1997). The qualitative method concentrates more on feelings, behavior and words than on numbers (Sharma, 2009). Qualitative research explores complex phenomena (Tong, Sainsbury & Craig, 2007). Denzin and Lincoln (2011) described that a qualitative research design is related to an *interpretive* philosophy. One of the drawbacks of the qualitative method is that it involves more time and skills to understand and analyze the data (Sharma, 2009). The qualitative data may be collected with a single technique, such as interviews. It can also be collected with more than one technique, such as interviews, focus group discussion and observation. A mixed research method is considered as an appropriate research method (Bryman, 1988) because this approach provides superior results as compared to other research methods (Johnson & Onwuegbuzie, 2004). Creswell, Klassen, Plano Clark and Smith (2011) stated that in quantitative methodology, the researcher uses a positivist philosophical strand which is primarily a deductive approach, whereas, in qualitative methods, the researcher uses an interpretivist philosophical which is mainly an inductive approach. Both qualitative and quantitative methods are respected by pragmatists, but the selection of a method depends on the nature of the study. Pragmatism supports a mixed methods research design (Saunders, et al., 2016). However, the mixed method is time-consuming because the researcher has to know more than one approaches and procedures to mix the results appropriately. Working with both quantitative and qualitative methods simultaneously is sometimes hard for a single researcher.

6.3.4 Fourth-Layer: Strategies

Saunders et al. (2016) classified a methodological choice layer into the following to collect data: (1) Survey, (2) Experiment, (3) Archival research, (4) Case study, (5) Ethnography, (6) Action research, (7) Grounded theory, and 8) Natural inquiry research. The first two research strategies (i.e. Experiment and survey) are entirely associated with quantitative research. The research strategies archival research and case study may contain quantitative or qualitative study or a mixed design combining both. The remaining four strategies, i.e. Ethnography, Action research, and Grounded theory, are strategies mainly linked to qualitative research design (Saunders et al., 2016 p. 178).

6.3.5 Fifth-Layer: Time Horizon

Saunders et al., (2009, 2016) classified the time horizon layer into two categories: longitudinal and cross-sectional. The longitudinal study is conducted over a prolonged period. The key strength of this research is its ability to study change and development over time. The cross-sectional study is conducted for one time only. For instance, the survey of the instructors or employees is one of the examples of a cross-sectional study (Robson, 2002; Saunders et al., 2016).

6.3.6 Sixth-Layer: Data Collection Techniques and Procedures

This step involves data collection and analysis procedures employed. For instance, data collection involves a survey questionnaire and the analysis includes exploratory factor analysis, descriptive analysis, confirmatory factor analysis, and structural path analysis.

6.3.7 Adopted Research Design and Methodology

In light of the discussion on the research process, it would be easy to understand the rationale of research design and philosophies adopted for this study. In this study, the explanatory technique is selected because in the context of Higher Education the variables (of UTAUT2) for the research were largely well known in the literature and statistical techniques (SPSS/Amos) are used to identify the significant variables. The selected research design for this study has been summarized in the following Table 6.1.

6.4 Sampling Techniques

Sampling is the procedure of choosing a few participants from a larger group to represent the entire group (Kumar, 2010). The sample size has a direct effect on the accuracy of the findings in the population (Burns & Bush, 1998). Representative data collection requires consideration of an appropriate sampling technique. The more the representativeness, the more the generalizability of the findings and, therefore, the better the quality of the research (Sarantakos, 2012). Sampling techniques are required to be understood by

Table 6.1 Summary of research design and methodology used for this study

Layer number	Layer name	Adopted approach
First-layer	Philosophy	Positivism
Second-layer	Research approach	Deductive
Third-layer	Research strategy	Quantitative
Fourth-layer	Methodology choice	Survey
Fifth-layer	Nature of research-time horizon	Cross-sectional
Sixth-layer	Data collection and analysis	–Techniques and procedures of quantitative analysis such as EFA, CFA

Source Saunders et al. (2016)

researchers to achieve the appropriate data. Selecting a correct sampling technique depends on the population, availability, and accessibility of the resources (Saunders et al., 2009). Probability sampling (representative) and non-probability sampling (judgmental) are two kinds of sampling techniques. A probability sampling guarantees that the entire population has an equal chance to be selected. Simple random, systematic, cluster, stratified random, and multi-stage are five techniques under probability sampling (Saunders et al., 2009, 2016). Non-Probability sampling is used when the entire population cannot be selected. Purposive, quota, self-selection, snowball, and convenience are the techniques under non-probability sampling (Saunders et al., 2009).

6.4.1 Target Population

The process starts with defining the target population and specifying the sampling frame. The target population is the main population of the research study, and it is a subset of the overall population (Saunders et al., 2016). The target population for this research study was the entire teaching staff of Saudi Higher Educational Institutions. However, this study is limited to the institutions of the Eastern Province of Saudi Arabia where the medium of instruction is English. The institutions include six universities and two community colleges (see the following Table 6.2). All emails sent through ITC-KFUPM reach all departments, including all community colleges under its umbrella. There are seven technical colleges, one military college, and one women's medical college in the Eastern region, but were not incorporated in the data collection because the medium of instruction in most of these colleges is Arabic. Furthermore, it was not possible to approach the women's colleges to get the data. Regarding the total number of instructors in the target academic institutions, the following information was collected through visiting their websites and contacting with the Human Resource department via phone and email.

6.4.2 Sampling Frame

A sampling frame is "the complete list of all the cases in the population, from which a probability sample is drawn" (Saunders et al., 2016, p.727). This study deals with the users (adopters) of the LMS. In an exploratory study, it was discovered that there are 58% adopters and 42% non-adopters of the LMS. The following Fig. 6.2 adopted from De Vaus (2002) shows moving from population to sample size.

6.4.3 Sample Size for Quantitative Data

In this study, the approximate target population size is 2005. A survey questionnaire was sent to the entire population of HEIs of the Eastern Province of Saudi Arabia. To calculate the sample size, the following procedures were adopted (without using standard deviation) as given in the following paragraphs:

Saunders et al. (2016), recommended the following steps for calculation of sample size proposed by De Vaus (2014):

$$\text{Step 1}: \quad n = \frac{z^2 \times (p) \times (1-p)}{c^2}$$

$$\text{Step 2}: \quad n' = \frac{n}{1 + \frac{n}{N}}$$

n   Minimum sample size = ?

z   confidence level (e.g. at 95% confidence level) = 1.96

p   population proportion (the percentage belonging to the specified category). From exploratory study, it was found that 40 out of 69 instructors are the adopters of the LMS. (see preliminary study) Therefore, the proportion of the users p = (40/69 × 100) = 58%

c   confidence interval (e.g. ± 5) = 5

N   Target population = 2005

Table 6.2 Target population

Institution name	No of instructors
King Fahd University of Petroleum and Minerals (KFUPM)	910
Dammam Community College (DCC)	40
Hafr Al-Batin Community College (HBCC)	37
Prince Mohammad University (PMU)	153
Dammam University (DU)	305
Jubail University College (JUC)	225
King Faisal University (KFU)	230
Al-Kharj University (KU)	105
Total number of instructors	2005

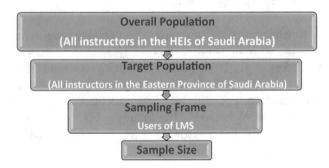

Fig. 6.2 Moving from Population to sample. *Source* De Vaus (2002)

n'   Adjusted minimum sample size = ?

Substituting these values in the formulae we get:

$$\text{Step 1}: \quad n = \frac{(1.96)^2 \times (58) \times (100 - 58)}{(5)^2} = 374.32$$

$$\text{Step 2}: \quad n' = \frac{374.32}{1 + \frac{374.32}{2005}} = 315.6$$

Hence, the approximate sample size for this study is **316**. The following Table (6.3) by Krejcie and Morgan (1970) indicates that the sample size for a total population of 2000, is 322.

Similarly, the sample size found for a total population of 2005, using online sampling calculator (available at http://www.surveysystem.com/sscalc.htm), online-calculator (available at http://fluidsurveys.com/survey-sample-size-calculator/) and, online-calculator (*available at* http://www.raosoft.com/samplesize.html) with a confidence level of 95% and a confidence interval of 5%, the sampling size is 323. Hence the researcher is confident about the sample size.

6.5   Quantitative Data Collection and Analysis

In this research, various primary and secondary sources were used for data collection. A survey questionnaire was the primary source of quantitative data collection. The review of the literature was a source of secondary data. Some second-hand data, such as data regarding the total number of instructors, and an approximate number of instructors using LMS was collected from different institutions. For instance, https://www.mohe.gov.sa/en/default.aspx, and http://www.citc.gov.sa were used for secondary data.

The survey instrument was divided into three sections and included questions on demography, questions on the core constructs of the UTAUT2 model and questions on the moderating variables of the model. According to Bell (1999) "the questionnaire is a widely used and useful instrument for collecting survey information providing structured, often numerical data, being able to be administered without the presence of the researcher, and often being comparatively straightforward to analyze" (p. 245). The questionnaire was designed with a five-point Likert scale from 1 (strongly disagree: lowest rating) to 5 (strongly agree: highest rating) to identify the adoption of LMS technology by instructors to get numeric results for statistical analysis. Five-point Likert scales were selected to offer the freedom to indicate the applicable rating for their situation. In most cases this scale has the ranges: strongly disagree (1), disagree (2), neutral (3), agree (4) and strongly agree (5). In this study, a survey instrument was adapted from Venkatesh, Morris, Davis and Davis (2003), Venkatesh, Thong and Xu 2012) and many other researchers who have used the UTAUT2 model in their

Table 6.3 Table for determining sample size

Table for determining sample size for a given population

N	S	N	S	N	S	N	S	N	S
10	10	100	80	280	160	800	260	2800	338
15	14	110	86	290	165	850	265	3000	341
20	19	120	92	300	169	900	269	3500	246
25	24	130	97	320	175	950	274	4000	351
30	28	140	103	340	181	1000	278	4500	351
35	32	150	108	360	186	1100	285	5000	357
40	36	160	113	380	181	1200	291	6000	361
45	40	180	118	400	196	1300	297	7000	364
50	44	190	123	420	201	1400	302	8000	367
55	48	200	127	440	205	1500	306	9000	368
60	52	210	132	460	210	1600	310	10000	373
65	56	220	136	480	214	1700	313	15000	375
70	59	230	140	500	217	1800	317	20000	377
75	63	240	144	550	225	1900	320	30000	379
80	66	250	148	600	234	2000	322	40000	380
85	70	260	152	650	242	2200	327	50000	381
90	73	270	155	700	248	2400	331	75000	382
95	76	270	159	750	256	2600	235	100000	384

Note "N" is population size
"S" is sample size
Source Krejcie and Morgan (1970)

research. Previously developed survey instruments are preferred for use because they carry reliability and validity tests (Henderson, Morris, Carol & Fitz-Gibbon, 1987). Furthermore, no appropriate and comprehensive questionnaire on culture could be found in the literature (Kiljander, 2004). The lack of suitable standardized questionnaires to research the influence of culture is a big problem (Renaud & Biljon, 2008). Therefore, the questions used on cultural dimensions were adapted from Hofstede's cultural dimensions (VSM 2008).

In the light of an exploratory study, two types of surveys were developed: one for *adopters* and another for *non-adopters* of LMS. To get better responses from surveys, both online and paper-based survey techniques were used.

Online Survey: In the online survey, based on their adopters or non-adopter category, the participants were moved automatically to their relevant survey questionnaire. The online survey on Google was made available at http://goo.gl/forms/TSXOQKFMvh for the instructors to fill their responses. The data collection through on-line format offers many *advantages* over other formats. It is an efficient and fast way to reach a relatively large number of instructors in various institutes. Internet surveys reduce time and costs substantially. Another advantage is that using an Internet survey, there is no need for 'data entry, and coding' because the data were entered by the participants and responses are

automatically saved in an electronic format (Sills & Song, 2002). One of the *limitations* pointed out by Dillman (2000), is the limited access to the computer or internet, varying degrees of computer literacy are common concerns of using online surveys. This limitation is not the barrier in this study because the sample includes instructors with the necessary computer skills, daily access to the Internet and e-mail.

Paper-based Survey: The researcher also used a personally administered, paper-based survey to increase the response rate. Bell (1993) argued that "*the research instrument is merely the tool to enable you to gather data, and it is important to select the best tool for the job*" (p. 66). This strategy works well and increases the response rate. The personalized email reminders were also sent to enhance the response rate. Some researchers argue that the data collection process is a challenging job. One of the main reasons is the resistance to surveys and interviews (Taylor & Todd, 1995) probably due to lack of time or lack of interest. Therefore, in the presence of the resistance, every possible effort must be made to improve the response rate.

Pilot Study: Many researchers suggest that a pilot study should be carried out before starting on the main data collection. The results and findings of the pilot study can be used to improve not only the survey instrument but also in improving the measures for sampling and data collection. According to Baker (1994), a pilot study is a trial study conducted on a small scale to pre-test the instrument in the preparation of the main study. Many serious errors can be avoided by taking the time to make an adequate pre-test. Dillman's (2000) suggestion was used as guidelines for the pilot study. First, the survey questionnaire was reviewed by the experts, researchers and knowledgeable colleagues to ensure its relevancy, completeness, and appropriateness. The *construct validity* was achieved by a discussion of survey instrument with research experts to check if the questions are appropriate for measuring the variables. The suggestions from the supervisor and the research experts were taken into consideration. The *content validity* was checked by a discussion on questions with a non-native instructor. An English instructor was requested to check the questionnaire for typographical errors. Lastly, a pilot study was conducted to get reliable and robust data that could lead the research towards reliable results.

6.6 Instrument Reliability

Evaluating the reliability of any survey instrument is an essential part of the analysis. Saunders et al. (2009, 2016) stated that reliability means that if the research is conducted over different periods, the results should be consistent. The reliability is *"the extent to which your data collection techniques or analyzing procedures will yield consistent findings"* (Saunders et al. 2009, p. 156). The *Cronbach alpha* reliability coefficient is used for measuring the *reliability* of the instrument (Im, Il Hong, Seongtae Kang, 2011). In this research, *Cronbach's alpha* is one of the main elements to evaluate the overall reliability of each variable, and the instrument as well. The value of Cronbach's alpha ranges from 0 to 1. According to George and Mallery (2003), as a rule of thumb, Alpha values should be more than 0.7. A value of Cronbach's alpha more than 0.7 is taken as a sufficient and reliable value (Schutte, Toppinnen, Kalimo & Schaufeli, 2000). Although the scales and items for the constructs of UTAUT2 were adapted from Venkatesh et al. (2003, 2012) that have already been tested and validated, it would be more appropriate to re-assess the reliability and validity of the instrument again because of its novel context (Nistor, Lerche, Weinberger, Ceobanu & Heymann, 2014).

6.7 Instrument Validity

According to Smith, Passmore and Faught (2009), validity is about not only the instrument itself but also about the outcome. Srivastava (2010) described that content validity, criterion validity, and construct validity are three types of empirical research validity. To achieve the validity, the six criteria that must be satisfied are highlighted by Messick

(1995) are that: (a) content is representative and relevant, (b) content is substantive and is supported by a proper theory, (c) the instrument structure must be the representative of the given construct, (d) outcomes should have generalizability strength (e) the outcome obtained should have the ability to be related to other external variables, and (f) the consequences and implications of the yielded outcome should have a significant and consequential impact. Although the questions adopted from Venkatesh et al. (2003, 2012) were validated by previous researchers, new items of the constructs must also be content-validated. For this research, to ensure that the instrument is constructed properly, instructors from various institutions such as KFUPM, DU, DCC, and other research experts having a thorough knowledge of technology adoption models were consulted to assess the content validity of the instrument.

6.8 Confidentiality and Ethics

After getting ethical approval from the researcher's institution, the survey link was sent to the Director of Information and Technology Centre (ITC) to distribute the survey questionnaire to all instructors. The higher management of other universities was also contacted to send the survey link to their instructors for filling the survey. In the survey, the instructors were given a briefing regarding the objectives and use of the research findings. The consent letter of voluntary participation was to assure that there would not be any physical or psychological harm to the participants and personal information would be kept confidential by the researcher. Ethical principles regarding Saudi society and the Islamic religion were also considered.

6.9 Summary

In this study, the Saunders et al. (2016)'s layered onion model was used to choose the most appropriate research method to answer the research questions and to meet the objectives of this study. In this chapter, the philosophy of research, methodology, and strategy to be used were discussed. Data collection, analysis procedures, and their justification are also discussed in this chapter. The strategies to improve the validity and to enhance the response rate were also discussed. This chapter also discussed how to enhance the reliability and validity of the study so that any possible bias could be minimized.

References

Baker, T. L. (1994). *Doing social research* (2nd ed). McGraw-Hill Inc.
Bell, J. (1993). *Doing your research project: A guide to first-time researchers in education and social science*. Milton Keynes: Open

University Press. https://books.google.com.sa/books?hl=en&lr=&id=Uo9FBgAAQBAJ&oi=fnd&pg=PR1&dq=Bell,+J.+(1993).+Doing+your+research+project:+A+guide+to+first-time+researchers+in+education+&ots=i_7zZ_ayDA&sig=Zz1TaqSUU4ahwoUNEQo1yqGUeEQ&redir_esc=y#v=onepage&q=Bell%2CJ.(1993).Doingyourresearchproject%3AAguidetofirst-timeresearchersineducation&f=false.

Bell, J. (1999). *Doing Your Research Project. A guide for first time researchers in education and social sciences.* Open University Press.

Bryman, A. (1988). *Quantity and quality in social research.* Unwin Hyman.

Burell, G., & Morgan, G. (1979). *Sociological paradigms and organisational analysis.* Heinemann.

Burns A. C. & Bush R. F. (1998). *Marketing research* (2nd ed.). Prentice Hall International, Inc.

Cavana, R. Y., Delahaye, B. L., & Sekaran, U. (2001). *Applied business research: Qualitative and quantitative methods.* New York: Wiley.

Chen, W., & Hirschheim, R. (2004). A paradigmatic and methodological examination of information systems research from 1991 to 2001. *Information Systems Journal, 14*(3), 197–235. https://doi.org/10.1111/j.1365-2575.2004.00173.x.

Collis, J., & Hussey, R. (1997). *Business research: A practical guide for undergraduate and postgraduate students.* Palgrave Macmillan.

Creswell, J. W. (2003). *Research design: Qualitative, quantitative, and mixed methods approaches.* SAGE Publications.

Creswell, J. W. (2009). *Research design: Qualitative, quantitative and mixed methods approach* (3rd ed.). SAGE.

Creswell, J.W., Klassen, A. C., Plano Clark, V. L., & Smith, K. C. (2011). *Best practices for mixed methods research in the health sciences.* Bethesda, MD: National Institutes of Health 10.

De Vaus, D. (2002). *Surveys in social research.* Routledge Taylor & Francis Group, London & Newyork.

De Vaus, D. A. (2014). *Surveys in social research* (6th ed.). Routledge.

Denzin, N.K., & Lincoln, Y. S. (2011). *The SAGE handbook of qualitative research.* Sage.

Dillman, D. A. (2000). *Mail and internet surveys: The tailored design method.* New York: Wiley.

Fishman, D. B. (1991). An introduction to the experimental versus the pragmatic paradigm in evaluation. *Evaluation and Program Planning, 14*(4), 353–363.

George, D., & Mallery, P. (2003). *SPSS for windows step by step: A simple guide and reference* (4th ed.). Allyn and Bacon.

Guba, E. (1990). *The paradigm dialogue.* Sage Publications.

Henderson, M. E., Morris, L. L., Carol, T., & Fitz-Gibbon. (1987). *How to measure attitudes.*

Hussey, J., & Hussey, R. (1997). *Business research: A practical guide for undergraduate and postgraduate students.* Palgrave Macmillan.

Im, Il Hong, & Seongtae Kang, M. S. (2011). An international comparison of technology adoption Testing the UTAUT model. *Information & Management, 48*(1), 1–8. https://doi.org/10.1016/j.im.2010.09.001.

Johnson, R. B., & Onwuegbuzie, A. (2004). Mixed methods research: A research paradigm whose time has come. *Educational Researcher, 33*(7), 14–26.

Kiljander, H. (2004). *Evolution and usability of mobile phone interaction styles.* Helsinki: Helsinki University of Technology.

Knudsen, C. (2002). *Pluralism, scientific progress and the structure of organization studies.* Handelshøjskolen i København: Institut for Industriøkonomi og Virksomhedsstrategi.

Krejcie, R., & Morgan, D. (1970). Determining sample size for research activities. *Educational and Psychological Measurement.*

Kumar, R. (2010). *Research methodology: A step-by-step guide for beginners* (3rd ed.). SAGE Publications Ltd.

Merriam, S. B. (1998). *Qualitative research and case study applications in education.* Jossey-Bass.

Messick, S. (1995). Validity of psychological assessment: Validation of inferences from persons' responses and performances as scientific inquiry into score meaning. *American Psychologist, 50*(9), 741–749. https://doi.org/10.1037/0003-066X.50.9.741.

Nistor, N., Lerche, T., Weinberger, A., Ceobanu, C., & Heymann, O. (2014). Towards the integration of culture into the unified theory of acceptance and use of technology. *British Journal of Educational Technology, 45*(1), 36–55. https://doi.org/10.1111/j.1467-8535.2012.01383.x.

Pansiri, J. (2005). Pragmatism: A methodological approach to researching strategic alliances in tourism. *2*(3), 191–206.

Parvaiz, G. S., Mufti, O., & Wahab, M. (2016). Pragmatism for mixed method research at higher education level. *8*(2), 67–78.

QiYing S., Adams, C. (2010). Consumers' attitudes toward Mobile Commerce. *International Journal of E-Services and Mobile Applications, 2*(1). https://doi.org/10.4018/jesma.2010101601.

Renaud, K., & Biljon, J. V. (2008). Predicting technology acceptance and adoption by the elderly: A qualitative study. *ACM.*

Robson, C. (2002). *Real world research: A resource for social scientists and practitioner-researchers* (2nd ed.). Blackwell Publishing.

Sarantakos, S. (2012). *Social research.* Palgrave Macmillan.

Saunders, M., Lewis, P., & Thornhill, A. (2009). *Research methods for business students* (5th ed.). Harlow: Pearson Education Limited.

Saunders, M., Lewis, P., & Thornhill, A. (2016). *Research methods for business students.* Harlow: Pearson Education. https://doi.org/10.1017/CBO9781107415324.004.

Schutte, N., Toppinnen, S., Kalimo, R., & Schaufeli, W. (2000). The factorial validity of the Maslach Burnout Inventory-General Survey across occupational groups and nations. *Journal of Occupational and Organizational Psychology, 73,* 53–66.

Sekaran, U., & Bougie, R. (2010). *Research methods for business: A skill building approach.* New York: Wiley.

Sharma, U. (2009). Qualitative research in business and management. *Qualitative Research in Accounting and Management, 6*(4), 292–296.

Sills, S. J., & Song, C. (2002). Innovations in survey research an application of web-based surveys. *Social Science Computer Review, 20*(1), 22–30.

Smith, G. G., Passmore, D., & Faught, T. (2009). The challenges of online nursing education. *The Internet and Higher Education, 12*(2), 98–103. https://doi.org/10.1016/j.iheduc.2009.06.007.

Srivastava, T. N. (2010). *Business research methodology.* Tata McGraw Hill Education.

Taylor, S., & Todd, P. (1995). Understanding information technology usage: A test of competing models. *Information Systems Research, 6*(4), 144–176.

Tong, A., Sainsbury, P., & Craig, J. (2007). Consolidated criterio for reporting qualitative research (COREQ): A 32- item checklist for interviews and focus group. *International Journal of Qualitative in Health Care, 19*(6), 349–357. https://doi.org/10.1093/intqhc/mzm042.

Venkatesh, V., Morris, M. G., Davis, G. B., & Davis, F. D. (2003). User acceptance of information technology: Toward a unified view. *27*(3), 425–478.

Venkatesh, V, Thong, J., & Xu, X. (2012). Consumer acceptance and user of information technology: Extending the unified theory of acceptance and use of technology. *MIS Quarterly, 36*(1), 157–178. http://ezproxy.library.capella.edu/login?; http://search.ebscohost.com/login.aspx?direct=true&db=iih&AN=71154941&site=ehost-live&scope=site.

Yates, L. (2004). *What does good education research look like?* Open University Press.

Yin, R. K. (2003). *Case study research: Design and methods* (3rd ed.). Sage Publications.

Empirical Evidence of LMS Adoption in the Middle East

7.1 Introduction

This chapter explains the process of data preparation, results, and preliminary data analysis. The prime focus of this chapter is the appropriateness of the data obtained about data analysis. This chapter presents the preliminary data results and the statistical methods applied in data analysis. In this chapter, the personal profile of the participants and the descriptive data analysis are discussed. The descriptive data analysis of the core variables (such as effort expectancy, performance expectancy, social influence, facilitating conditions, hedonic motivation, and habit) and moderating variables (experience, age, and cultural dimensions) is discussed. In the last section, the reliability, correlation, factor analysis, and regression analysis are discussed.

7.2 Descriptive Statistics

This section provides a descriptive analysis of the participants. The main attributes of the personal profile of the participants include nationality, age, years of LMS experience, department, teaching rank of instructors, and educational level. The participants for this study include the Arab and non-Arab instructors from various educational institutions in the Eastern Province of Saudi Arabia. The survey questionnaire was sent to the population of 2005 instructors. The number of received responses was 396. However, 22 responses were incomplete, wrongly filled surveys, and the participants were not the teaching staff; therefore, those responses were discarded. Hence, the number of valid responses was 374, as shown in Table 7.1. After data entry, an SPSS 22/AMOS statistical package was used for the analyses, such as descriptive analysis, Cronbach's alpha test, and factor analysis. The entire population of this study consisted of male instructors. Therefore, the 'gender' data were removed from the research. If in the future, male and female instructors teach in Saudi universities, then gender data will be added to the study.

7.3 Quantitative Data Collection and Analysis

The steps involved in the data analysis are given in Fig. 7.1.

Step 1: Investigation for Potential Biases

One of the most important steps in starting data analysis was checking the database for potential biases. The collected data were inspected for non-response bias and a common method bias.

- *Non-response Bias*: This bias relates to the probability that the participants who responded differ from those who did not, which does not allow to conclude the whole sample. As a result, the item non-response provides incomplete information and affects the reliability of the results. To avoid non-responses bias, one of the solutions suggested by the literature is the reduction of non-response itself.
- *Common Method Bias*: The collected data from the same instructor with the same instrument can be a cause of common method bias. When more variables or constructs of the same survey questionnaire are measured from the same instructor, there are chances that participants may respond in the same fashion and in a similar direction. Subsequently, the two constructs may correlate with each other and may lead to a wrong conclusion. According to Podsakoff, MacKenzie, Lee, & Podsakoff (2003) and Podsakoff & Organ (1986), this kind of bias impacts the research results and needs to be controlled.

In this study, the researcher used *Harman's single-factor test,* to address this issue, as recommended by Podsakoff et al. (2003). In this regard, the factor analysis (EFA) was conducted using SPSS with the options: in the extraction process, use several factors to extract '1'; in the rotation method, the selected method was '*none*'. After running this test, one factor emerged that explains 32.4% of the variance,

© Springer Nature Switzerland AG 2021
R. A. Khan and H. Qudrat-Ullah, *Adoption of LMS in Higher Educational Institutions of the Middle East,*
Advances in Science, Technology & Innovation, https://doi.org/10.1007/978-3-030-50112-9_7

Table 7.1 Descriptive statistics

Category (Nationality)	Frequency	Percent
Saudi Nationals	53	14.17%
Other Arabs (Non-Saudi)	65	17.38%
Non-Arabs (Non-Saudi)	265	68.45%
Total	**374**	

Category (Age)	Frequency	Percent
Under 30 years	28	7. 49%
31 to 35 years	53	14.17%
36 to 40 years	52	13.90%
41 to 45 years	77	20.59%
46 to 50 years	62	16.58%
51 to 55 years	42	11.23%
More than 55 years	60	16.04%
Total	**374**	

Category (Educational Level)	Frequency	Percent
PhD	180	48.13%
MPhil	9	2.41%
Master's Degree	145	38.77%
Bachelor's Degree	40	10.70%
Total	**374**	

Category (Teaching and Non-Teaching Staff)	Frequency	Percent
Teaching instructors	357*	95%
Non-teaching instructors	17	5%
Total	**374**	

* Only Teaching Faculty Members were included in the data analysis

Category (Educational Institutions)	Frequency	Percent
King Fahd University of Petroleum and Minerals (KFUPM)	150	39.06%
Dammam University (DU)	94	24.48%
King Faisal University (KFU)	32	8.33%
Jubail University College (JUC)	40	10.42%
Prince Mohammad University (PMU)	18	4.69%
Haffar Al Batin Community College (HBCC)	19	4.95%
Dammam Community College (DCC)	21	5.47%
Al-Kharage University (KU)*	0	0
Total	**374**	

Category (Technological Awareness)	Frequency	Percent
Not Aware	49	13%
Low Awareness	162	40%
High Awareness	163	43%
Total	**374**	

Category (*Adopters and Non-Adopters*)	Frequency	Percent
Users of LMS	310	82.88%
Non-users of LMS	64	17.11%
Total	**374**	

NOTE: The entire statistical data analysis is based on 310 'adopters' of the LMS (i.e., users of LMS). Hence, 64 'non-adopters' (non-users) of LMS are not included in quantitative analysis.

Category of users with LMS experience	Frequency	Percent
< 1 year (new users)	64	20.65%
1 to 3 years	126	40.65%
4 to 6 years	82	26.45%
7 to 9 years	18	5.81%
>10 years	20	6.45%
Total	**310***	

* Only 310 adopters of LMS are used in the data analysis.

Category of users (LMS usage)	Frequency	Percent
1 to 10 Hrs.	248	80%
11 to 20 Hrs.	41	13%
More than 20 hrs.	21	7%
Total	**310**	

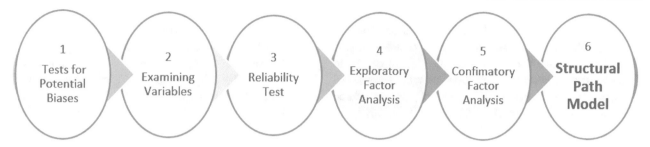

Fig. 7.1 Structure of data analysis

which is an acceptable range. Anonymity also guaranteed to lessen social desirability effects, which are thought to be a basis of 'common method variance'. *In this study, common method bias was not found in the data and thus was not being a threat to the study results.*

Step 2: Investigation of Variables

All variables were prudently examined for healthier understudying of their characteristics and connecting to the subsequent stage data analysis, such as distributions, and correlations.

- **Data Cleaning and Data Screening**: After confirmation that no bias was present, the next stage was to check all items individually that involve data cleaning and data screening. A total of 2005 surveys along with consent letters were distributed (via an email link and paper-based) to the instructors, and 396 surveys were received. After gathering the surveys, the acquired data were entered into the AMOS/SPSS software.
- **Handling Outliers**: According to Pallant (2013), an important phase in data screening is to detect out-of-range values. This step was performed by focusing on outliers. When some data is found to be different from the majority of the other responses, this data is known as *outliers*. This problem affects the conclusion of the study and could change the standard deviation and impacts the normality of the data (Field, 2009; Osborne & Overbay, 2004). All real data contain outliers (Ritter & Gallegos, 1997). In this research, most of the variables are measured on a five-point scale (Likert scale) ranging from strongly disagree to strongly agree (1 to 5). Therefore, the threat of outliers is not of concern because all values range between 1 and 5, in which case the extreme values (1 and 5) are the legitimate outliers (Osborne & Overbay 2004). Another possible reason for an outlier is the human error in data entry. Since the data is imported from the online survey into the statistical package, there is no human intervention. In this study, the responses were tested for outliers through boxplot using SPSS and some outliers were noticed. Hence, the only outliers present in this data are the ones which may be considered as legitimate outliers and do not pose any real threat.

- **Missing Data**: The next stage included the inspection of data for missing values. Twenty-two out of 396 responses were unusable due to incomplete and missing responses and were immediately discarded, yielding 374 responses leaving a response rate of 16.10%. However, small missing values were exchanged with the median, which is a suitable technique given the distribution properties of the data and the relatively low number of missing values.
- **Normality**: It is imperative to identify any non-normal distributions that might threaten the validity of the collected data. To evaluate normality, *skewness and kurtosis* are two main recommended tests and is normally indicated by a bell-shape symmetrical curve (Field, 2009). The *Skewness* provides information about the symmetry of the data distribution. In simple words, skewness indicates that a variable skewed when its meaning is not in the center of the distribution. The *Kurtosis* gives information about the 'peakedness' of the distribution. A distribution is a normal distribution when kurtosis and skewness values are equal to zero (Azzalini & Capitanio, 1999). When skewness and kurtosis values are zero, then the data is considered to be distributed normally, but if any value is increased positively or negatively, then normality is decreased (Tabachnick & Fidell, 2007). Several authors have supported that if the absolute value of kurtosis is less than 10 and the absolute value of skewness is less than 3, data is considered to be distributed normally (Hair, Black, Babin & Anderson, 2010). The negative skewness shows that the distribution is skewed to the left side.

In this research study, all of the distribution values were found to be normal because the absolute values of kurtosis and skewness were below 3 and 2, respectively. Thus, no indication of non-normal distributions is noticed. The results of this study revealed an acceptable range of skewness (under 2) and kurtosis (under 8).

Step 3: Test for Reliability

This step involves mainly the validity and reliability of the constructs as described below:

- **Construct Validity and Reliability**: As the survey was adapted from the UTAUT2 model, it was necessary to reassess their reliability and validity in the context of higher educational institutions. The reliability is the degree to which data is error-free and therefore returns consistent results. It can be inspected in terms of consistency and stability of the results (McCullough & Best, 1979). In general, three tests are used for assessing the reliability: test-retest, internal-consistency, and alternative-forms (Peter, 1979). In the alternative-forms and test-retest, items of the research must be administered to the same participants at different times. To assess the internal-consistency, *split-half reliability methods* and *Cronbach's coefficient alpha methods* are used. However, the coefficient alpha is preferred over the split-half model because the results of the split-half model become unreliable when the numbers of items cannot be divided into two halves (Green & Salkind, 2005). According to Peter (1979), Cronbach's coefficient alpha is the most common and preferred method for evaluating the reliability of measures. The coefficients of 0.7 or above are generally considered an excellent value for reliability. George & Mallery (2003) developed a rule to interpret the Cronbach's coefficient alpha as: <0.5 is unacceptable, >0.5 stands poor, >0.6 stands questionable, >0.7 stands acceptable, >0.8 is good, and alpha > 0.9 is an excellent value.

Step 4: Exploratory Factor Analysis (EFA)

EFA is a multi-variate statistical method that has various uses, such as reduction of a great number of factors into a small number of factors, establishing main dimensions between latent and measured constructs, and offering the evidence of *construct validity* of scales (Williams, Brown & Onsman, 2012). In this study, the steps involved in the exploratory factor analysis (EFA) are shown in Figure 7.2.

a. **Data suitability**: The following are the recommended tests for data suitability.

- **Sample size** is vital in 'factor analysis'. Regarding sample size, different opinions are cited in the literature. For instance, Tabachnick & Fidell (2007) suggest a sample of at least 300, whereas Hair et al. (2010) recommend a sample size of 100 or greater. Williams et al. (2012) cited guidelines from different studies that sample size of 100 is poor, 200 is fair, and 300–500 is a very good sample size. The sample size of 1000 or greater is an excellent data for analysis. *In this research, the number of usable responses was (310) appropriate for the analysis.*

- **Correlation matrix**: In EFA, a correlation matrix reflects the relationships between individual variables. Tabachnick & Fidell (2007) suggested examining the correlation matrix for correlation coefficients over 0.30. If the correlation is less than 0.30, then the researcher should retest the appropriateness of the factor analysis (multi-collinearity). *In this study, the correlation coefficients were in an acceptable range.*

- **Kaiser-Meyer-Olkin (KMO)**: Several tests are performed to evaluate the suitability of the data before the extraction of the factors such as Kaiser-Meyer-Olkin (KMO), measure of sampling adequacy, and Bartlett's test of sphericity (Tabachnick & Fidell, 2007). The KMO index varies from 0 to 1. The KMO value close to 0 indicates the diffusion in the correlation. The KMO value of 0.50 is suitable for factor analysis, whereas a value close to 1 ensures a compact correlation pattern showing the factor analysis is suitable. In this study, the items having loading less than 0.40 were omitted from the analysis as they were treated to be weak items (Hair et al., 2010). The KMO value of current research is 0.894, which confirms the suitability of the factor analysis, with a significant value using Bartlett's test of sphericity of 0.000 at p < 0.01 (Tabachnick & Fidell, 2007), as shown in Table 7.2.

b. **Extraction of factors:** The purpose of the rotation is to reduce the factor structure of items. In factor analysis, there are several means to extract factors such as principal axis factoring, principal components analysis, image factoring, maximum likelihood, alpha factoring, and canonical. However, the principal components analysis and principal axis factoring are the most commonly used methods.

c. **Criteria of factor extraction:** The objective of the data extraction is to reduce a large number of items into related variables or factors. The majority of the researchers usually use multiple criteria (Hair et al., 2010), such as cumulative percentage of variance and eigenvalue >1 rule, and the Scree test. The researchers suggest applying multiple approaches for factor extraction.

Fig. 7.2 Five steps in exploratory factor analysis. Adapted from Williams et al. (2012)

Table 7.2 KMO and Bartlett's test (SPSS Output)

KMO and Bartlett's Test		
Kaiser-Meyer-Olkin measure of sampling adequacy		0.894
Bartlett's test of sphericity	Approx. Chi-Square	6466.977
	Df	465
	Sig	0.000

Kaiser-Meyer-Olkin (KMO) measure of sampling adequacy and Bartlett's test of sphericity

- **Cumulative Percentage of Variance and Eigen-value >1 Rule**: The eigenvalue explains the degree of variation explained by a factor. The eigenvalue of 1 represents an extensive amount of variation (Field, 2009). There is no fixed threshold, although the researchers have recommended certain percentages. For instance, Hair et al. (2010) describe that in natural sciences, the explained variance should be at least 95% and in the humanities, the explained variance can be as low as 50–60%.

  Figure 7.3 shows the Scree plot for eigenvalue and components for this study.

- **Scree test**: The name was given by Cattell (1966) due to visual similarities to scree (rock debris) at the base of a mountain. The Scree plot is another disagreement area and debate for the researchers where the interpretation of Scree plots depends on the researchers' judgment (Tabachnick & Fidell, 2007).

d. **Rotational method selection**: Rotation produces a more interpretable and simplified solution by minimizing the low-item loadings and maximizing the high-item loadings. There are two famous rotation methods. The *orthogonal varimax rotation method* by Thompson (2004) is a popular rotational method and used in this study, which yields factor structures that are uncorrelated (Williams et al., 2012), whereas *the oblique rotation technique* yields the factors that are correlated. In this study, the method used was based on

the principal components factoring method with varimax rotation on the correlations of the observed variables.

e. **Interpretation**: It involves the examination of the researcher to give a name or theme to the factors. For instance, a construct may include four variables, which are all related to the comfort of the use of technology; therefore, the researcher created a name of this theme as 'ease of use' for that construct. In the context of this study, all items of the survey conducted were divided into eight groups, where every group is known as a 'construct' or a 'factor' and is labeled based on the type of items that have made it. For instance, group 1 in the following table is made of items regarding the usefulness and benefit (PE1 to PE5) of the LMS technology, and therefore it is labeled as 'performance expectancy', which is one of the constructs of UTAUT2. In group 2, items EE1 to EE5 are related to ease of use, and their group is labeled as effort expectancy, one of the constructs of UTAUT2. A similar approach has been adopted for labeling other groups. The names of all constructed have been adopted from the UTAUT2 model.

7.4   Overall Exploratory Factor Analysis

All items of the survey conducted were entered and factor analysis was run. The EFA produced *eight constructs*. These constructs were labeled in the light of literature related to

Fig. 7.3 Scree plot (researcher's SPSS output)

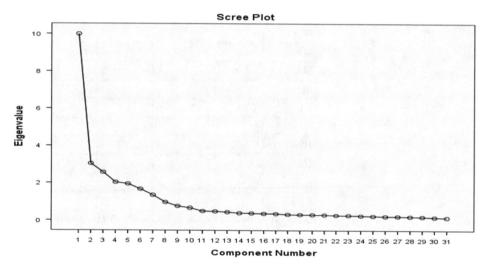

UTAUT2. The resulting model has KMO values of 0.894, whereas Tabachnick & Fidell's (2007) test of sphericity was significant (6466.97, p < 0.01). Principal component analysis (PCA) indicated eight components with eigenvalues exceeding one, explaining 75.95% of the variance. Therefore, the researcher is confident about the appropriateness of factor analysis.

Table 7.3 shows the results from the EFA of all constructs.

Table 7.3 shows the factor analysis of 31 items with eight constructs. The highest eigenvalue is 9.995, the total variance explained is 75.75%, and Cronbach's alpha is 0.922.

The correlations among all items were also inspected. The correlation matrix shows that most variables are correlated with each other. However, Cohen's (1977) provided criteria that correlations of 0.1 are taken as small, 0.3 is medium, and 0.5 is large and reflects that healthy correlations are observed among items from the same scale. After running an

Table 7.3 Component matrix for the overall questionnaire–factor analysis

Rotated component matrix

Items	Overall components of UTAUT2 model							
	1	2	3	4	5	6	7	8
BI1							0.684	
BI2							0.697	
BI3							0.727	
PE1		0.832						
PE2		0.873						
PE3		0.89						
PE4		0.812						
PE5		0.807						
EE1	0.821							
EE2	0.83							
EE3	0.829							
EE4	0.787							
EE5	0.824							
SI1			0.83					
SI2			0.826					
SI3			0.773					
SI4			0.832					
SI5			0.829					
FC1				0.802				
FC2				0.794				
FC3				0.797				
FC4				0.817				
M1					0.813			
M2					0.881			
M3					0.875			
H1						0.702		
H2						0.676		
H3						0.751		
UB1								0.831
UB2								0.88
UB3								0.897

Extraction method: principal component analysis, rotation method: varimax with Kaiser normalization, rotation converged in six iterations

EFA test, the next step is the model testing process. This step involves running a confirmatory analysis (EFA) test followed by a structural path testing.

7.5 Model Testing Procedures

Anderson & Gerbing (1988) classified the model testing process into two steps: confirmatory factor analysis and structural path model. In structural equation modeling (SEM), the CFA tests are considered as the critical starting (Hair et al., 2010). In this two-step method, a measurement model is estimated and then tested for the full structure model, as suggested by Townsend, Yeniyurt, Deligonul and Cavusgil (2004), Zou and Cavusgil (2002).

7.5.1 Confirmatory Factor Analysis (CFA)

This section provides the key findings regarding the initial measurement model fit. Using the results of the EFA stage, the confirmatory factor analysis (CFA) measurement model was developed and tested. The evaluation of the measurement model of CFA helps for a better understanding of how good the measurement items reflect the latent variables. In this study, the item loadings and fit-indices of all constructs are measured and are summarized in the following paragraphs: The fit statistics of the constructs of a model indicate a good fit to the data (Fındık & Özkan, 2010). Table 7.4 contains a summary of achieved fit indices for the overall model.

Many researchers agree on the following indices that should be reported for a good fit model (Fındık & Özkan, 2010):

- The primary fit indicator is the χ^2/df (chi-square/degrees of freedom ratio). The fit statistics of all constructs show a good fit to the data, where χ^2/df = 1.329, which does not exceed the recommended threshold. The lesser value is better, however, between 2 and 5. Therefore, the researcher considered other important indexes (Blunch, 2013). These indices include the root-mean-square residual (RMR) (Jöreskog & Sörbom, 1981), the root-mean-square error of approximation (RMSEA) (Steiger, 1990), the normed fit index (NFI) (Bentler and Bonnett, 1980), and the comparative fit index (CFI) (Bentler, 1990).

- **Root mean square error of approximation (RMSEA)** indexes for the overall model are equal to 0.033. The RMSEA value (0.33) meets the standards of <0.05. The RMSEA value <0.05 indicates good fit; a value <0.10 shows an acceptable fit and RMSEA >0.10 indicates a poor fit.
- **CFI** equals 0.979, which is above the recommended 0.90.
- **Normed fit index (NFI)** is 0.921, which is appropriate.
- The NFI value close to 0.95 reflects a good fit. NFI varies from 0 (no fit) to 1 (perfect fit).
- **Root mean square residual (RMR)** of 0.035 is a good fit because the lesser the RMR, the better the model, and RMR of zero shows a perfect fit.

The summary of path coefficients, regression weights, critical ratio (CR), and significance (P-values) for the constructs of the overall CFA model is shown in Table 7.5.

Reliability and validity of CFA model: One of the advantages of CFA is that while performing CFA testing, the reliability and validity of the variable can also be tested (Byrne, 2010). The higher is the reliability, the lower will be the measurement errors (Hair et al., 2010). The convergent validity is the extent to which the items *within a construct* are correlated. It indicates how well the items explain the factor. The convergent validity is evaluated by testing the modification coefficients, factor loadings, and variances. The discriminant validity is the extent to which the factors or constructs of a model are unique (Bagozzi & Yi, 1991; Hussain, Khan, & Al-Aomar, 2016). Using standardized regression weights and SPSS/AMOS's correlations into validity testing tools within the 'Stats Tools Package' (Gaskin, 2012), the validity and reliability testing results were calculated.

- **Average variance extracted (AVE):** This is a measure of convergent validity and should be above 0.5 (Hair et al., 2010). All AVE is above 0.5 except H-construct value, which is 0.305. The items belonging to the factor itself should better explain it than the items belonging to other factors (Straub, Boudreau & Gefen, 2004). It should be higher than MSV and ASV.
- **Composite reliability (CR):** It measures the reliability of the factors and should ideally be above 0.75. All CR values are above 0.75 except H-construct value.
- **Average shared squared variance (ASV):** ASV is similar to MSV, but takes the average of the squared

Table 7.4 Achieved fit indices for all constructs

Achieved fit indices for the overall model								
X2/df	GFI	AGFI	NFI	CFI	TLI	PCLOSE	RMSEA	RMR
1.329	0.905	0.883	0.921	0.979	0.975	1.000	0.033	0.035

Table 7.5 Overall CFA—summary of path coefficient of questionnaire items

Paths			Regression Weights		S.E.	C.R.	P-Value
			Estimate	Stand Estimate			
PE5	<---	PE	1.000	.785			
PE4	<---	PE	1.005	.734	.065	15.453	***
PE3	<---	PE	1.247	.829	.078	15.896	***
PE2	<---	PE	1.236	.897	.071	17.469	***
PE1	<---	PE	1.055	.863	.063	16.705	***
EE5	<---	EE	1.000	.828			
EE4	<---	EE	.998	.832	.057	17.462	***
EE3	<---	EE	1.093	.856	.060	18.227	***
EE2	<---	EE	1.108	.816	.065	16.962	***
EE1	<---	EE	1.135	.856	.062	18.212	***
SI5	<---	SI	1.000	.808			
SI4	<---	SI	.904	.773	.052	17.291	***
SI3	<---	SI	.938	.787	.062	15.226	***
SI2	<---	SI	1.009	.854	.060	16.934	***
SI1	<---	SI	1.061	.865	.062	17.196	***
FC4	<---	FC	1.000	.773			
FC3	<---	FC	.989	.766	.063	15.606	***
FC2	<---	FC	1.073	.839	.072	14.966	***
FC1	<---	FC	1.102	.857	.072	15.219	***
M3	<---	M	1.000	.857			
M2	<---	M	.925	.891	.049	19.009	***
M1	<---	M	.840	.834	.048	17.662	***
BI3	<---	BI	1.000	.830			
BI2	<---	BI	1.101	.867	.063	17.553	***
BI1	<---	BI	1.067	.818	.065	16.354	***
UB1	<---	UB	1.000	.841			
UB2	<---	UB	1.032	.807	.063	16.399	***
UB3	<---	UB	1.101	.903	.061	18.189	***
H3	<---	H	1.000	.558			
H2	<---	H	.929	.474	.183	5.076	***
H1	<---	H	1.160	.615	.216	5.379	***

Source SPSS/AMOS output

variances. It shows how much on an average is explained by items of other factors.

- **Maximum Shared Squared Variance (MSV):** The MSV in the model indicates how well is the factor explained by items outside the factor (i.e. items of other constructs).

It was noted that the CR for the habit construct (H) is <0.70; the AVE for H is also <0.50. Hence, there were validity concerns with the construct 'habit' (H). To achieve a better model fit, the construct habit (H) was dropped from the model and the test was repeated. After dropping H, the revised model produced a good model fit with (χ^2/df) = 1.29 and the RMSEA = 0.031.

7.5.2 Structural Path Model

The structural path model in the next stage provides the causal relations between independent variables and the dependent variables. Structural equation modeling has been extensively used in various fields, such as consumer behavior and marketing research because it offers many advantages over other procedures. SEM can investigate and correct unreliable measures when numerous indicators of each factor are available. The main benefit of using SEM is that it can examine the comprehensive theoretical frameworks in which the influences of factors are established across multiple levels of variables (Baumgartner & Homburg, 1996). The SEM is a way of a multi-variate statistical procedure that can assess the underlying latent constructs which are identified by factor analysis (Klem, 2000). In this study, from a measurement CFA, a structural path model was developed and tested with the maximum likelihood procedure in SPSS/AMOS. Also, an attempt was made to accommodate the influence of moderating variables (such as age, experience, and culture) on behavioral intention (BI) and use behavior (UB). In this model, the χ^2 is 418.585 with 328 degrees of freedom. The χ^2/df ratio is equal to 1.27. The other fit indexes are, respectively, RMSEA = 0.030, RMR = 0.031, CFI = 0.985, NFI = 0.935, RFI = 0.926 and other indicators that indicate a good fit. The final structural path model is shown in Figure 7.4.

The results of Figure 7.4 are summarized in Table 7.6. The table depicts the summary of path coefficients with their regression weights, SE, CR, and significant (p) values.

It is evident from Table 7.6 that the standardized regression coefficient (SRC) from FC to UB is 0.034 (with CR = 0.447 and p = 0.655) is a non-significant link. Therefore, the link FC → UB was removed from the final structural model and the model was rerun as recommended by Venkatesh, Thong, & Xu (2012).

It is important to note that the structural model (without considering the moderation impact) showed that R square for BI is 0.58 that explains 58% of the model, whereas R square of the UB is 0.235 that explains the 23.5% of the model. In other words, five independent variables explained 58% of the variance in the instructors' behavioral intentions to adopt the LMS in their face-to-face teaching.

Hence, the results from Figure 7.4 and path coefficient table of the final structural path model are summarized as:

1. **Path PE to BI**: The standardized regression coefficient (SRC) from PE to BI is 0.246 CR = 4.71, and p<0.01, which indicates a positive and significant relationship.
2. **Path EE to BI**: The SRC from EE to BI equals 0.198 with CR = 3.340 and p<0.01, which indicates a positive and significant relationship between EE and BI.
3. **Path SI to BI**: The SRC from SI to BI equals 0.144 with CR = 2.567 and p<0.01, which indicates a positive and significant relationship between SI and BI.
4. **Path FC to BI**: The SRC from FC to BI equals 0.270 with CR = 4.338 and p<0.01, which indicates a positive and significant link between FC and BI.
5. **Path M to BI**: The SRC from M to Bi is 0.214 with CR = 3.929 and p<0.01, which indicates a positive and significant link.

6. **Path BI to UB**: The SRC from BI to UB is 0.459 with CR = 5.68 and p<0.01, which indicates a positive and a significant relationship between BI and UB.

7.6 Tests for Moderation

In this study, age, experience, awareness, and cultural dimensions are treated as a moderator between independent variables (PE, EE, SI, FC, HM, and H) and dependent variable (BI). The procedure for applying moderation tests is adapted from Im, Il Hong & Seongtae Kang, (2011) and Gaskin (2012). In this study, the responses were collected in more than one category. To calculate the moderating effect in the SPSS/AMOS program, the first median of all moderators was calculated and then based on their median, all categories were converted into two groups. For instance, the responses for age were collected into seven categories, ranging from under 30 years to 60 years and above. Based on their median, all seven categories of age were converted into two groups, named as older age and younger age groups. A similar approach was applied to convert all other moderators into two groups such as low experience and high experience. In this test, the z value indicates the significance of the mean difference between the two groups (Byrne, 2010). The moderation testing process includes two main tests: (a) model-level moderation test and (b) path-by-path moderation test.

7.6.1 Model-level Moderation

In the moderation test, the model is run for two groups (low and high) without any constraints (i.e. free model) and the chi-square (χ^2) is noted. Then the model is constrained by setting the 'equality constraint' for the same groups and the model is run again. The chi-square (χ^2) is again noted. The chi-square (χ^2) values of both models are compared. Using the '*Stats Tools Package*' of Gaskin (2012), the chi-square (χ^2) values, along with their degrees of freedom (df), and the p-values are calculated for overall moderation (see Stats Wiki and Stats Tools Package (http://statwiki.kolobkreations.com/index.php?title=Main_Page)). The results of the moderation test at model-level are summarized in Table 7.7.

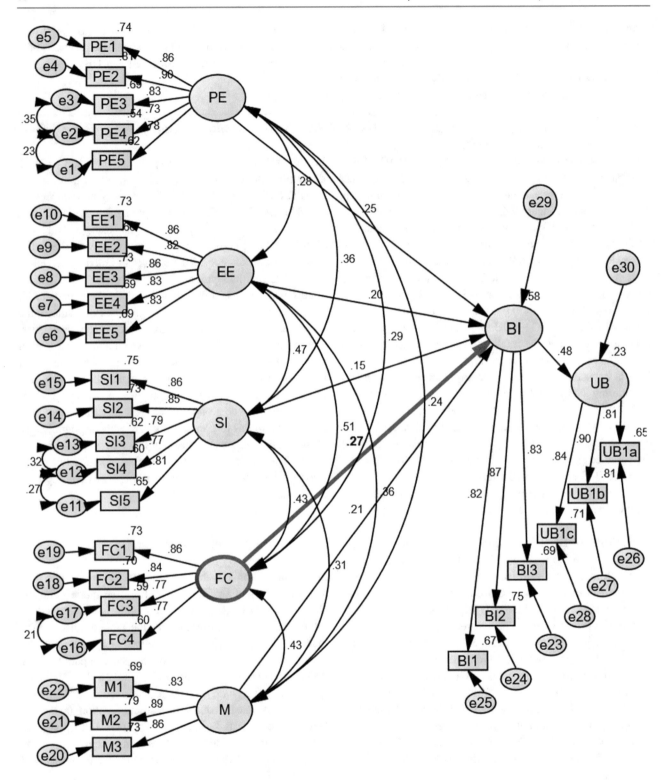

Fig. 7.4 Model 3—final structural path model. *Source* Drawing by SPSS/AMOS

It is evident from Table 7.7 that at model-level, experience, awareness, power distance, and uncertainty avoidance have a significant moderating impact on the dependent variable. The age, individualism, and masculinity have an insignificant moderating impact on the dependent variable.

Table 7.6 Summary of path coefficient of the constructs—path analysis

Paths			Regression Weights		S.E.	C.R.	P-Value
			Estimate	Stand Estimate			
BI	←	PE	.259	.246	.055	4.711	***
BI	←	EE	.188	.198	.056	3.340	***
BI	←	SI	.144	.148	.056	2.569	.010
BI	←	FC	.268	.270	.062	4.292	***
BI	←	M	.169	.214	.043	3.926	***
UB	**←**	**FC**	**.035**	**.034**	**.078**	**.447**	**.655**
UB	←	BI	.470	.457	.083	5.687	***

Table 7.7 Moderation test at model-level

Moderation Test at Model-Level

Dimension	χ^2/d		Unconstrained	Fully constrained	Difference	Number of groups	p-value	Invariant	Interpretation
Age	Chi-square		1078.572	1091.427	12.855	2	0.169	Yes	No moderation exists
	Df		808	817	2				
Experience	Chi-square		1042.94	1062.973	20.033	2	0.018	No	Groups are different at the model level
	Df		808	817	9				
Awareness	Chi-square		757.578	769.616	12.038	2	0.099	No	Groups are different at the model level
	Df		656	663	7				
Power Distance	Chi-square		783.844	800.461	16.617	2	0.011	No	Groups are different at the model level
	Df		656	662	6				
Uncertainty Avoidance	Chi-square		1010.478	1038.346	27.868	2	0.001	No	Groups are different at the model level
	Df		808	817	9				
Individualism	Chi-square		988.291	996.182	7.891	2	0.639	Yes	No moderation exists
	Df		808	818	10				
Masculinity	Chi-square		989.456	999.185	9.729	2	0.373	Yes	No moderation exists
	Df		808	817	9				

7.6.2 Path-by-Path Moderation

In the second step, path-by-path moderation is examined to check which paths are being moderated on an individual level. Path-by-path moderation is calculated by using 'regression weights' of groups and 'critical ratios' for the differences between the parameters (group differences), using tools within the 'Stats Tools Package' (see Stats Wiki and Stats Tools Package http://statwiki.kolobkreations.com/index.php?title=Main_Page). The estimates, p-values, and z-scores are calculated for each group and every path. Following the same procedure, path-by-path tests for other variables were run. However, path-by-path tests were executed only for the paths that are moderated by the moderating variables.

- **Path-by-Path moderation of experience**: Data were collected in different groups of experience (high experience and low experience) and a comparison was made. In this test, two paths were found to be moderated, that is, FC → BI and FC → UB.

  – FC → BI: The experience has a moderating impact between FC → BI (t = 1.661*, p<0.1) such that for high experience this relationship (0.374 at p<0.01) is *significant*; also for low experience this relationship is *significant* (0.155 at p<0.05). In other words, the relationship between FC → BI increases and is stronger (almost triple) for high-experience instructors.
  – FC → UB: There is a moderating effect of experience between FC → UB (t = 2.8***, p<0.01) such that for

high experience this relationship (–0.259 at p<0.05) is *negatively significant*, but for low-experience this relationship is *significant* (0.187 at p<0.01).

- **Path-by-path moderation of awareness**: In the path-by-path moderation test, the following moderating effects were observed.

  - EE → BI: There is a moderating effect of awareness between EE and BI (t = 2.04**, p<0.05). At high awareness this relationship (0.035 at p = 0.703) is insignificant. However, for low awareness this relationship is significant (0.272 at p = 0.000). This result shows that the instructors having low awareness give more importance to ease of use of the technology. However, ease of use becomes irrelevant at high awareness.
  - SI → BI: There is a moderating effect of awareness between SI and BI (t = 2.774***, p<0.01). The result shows that awareness interacts (moderates) the relationship between SI and BI differently. At high awareness, this relationship (t = 0.330 at p = 0.000) is significant. At low awareness, this relationship is insignificant (t =0.008 at p = 0.8977).
  - HM → BI: There is a moderating effect of awareness between HM and BI (t = 1.666*, p<0.10). At high awareness this relationship (0.228 at p = 0.000) is significant. However, for low awareness, this relationship is mildly significant (0.088 at p = 0.109). This result shows that the instructors with high awareness have a stronger relationship between HM and BI. In other words, high awareness influences their motivational level of the instructors to use LMS in their teaching.

- **Path-by-Path tests for cultural dimensions (PD, UA, IND, and MAS)**
- Responses for power distance (PD) dimensions were collected on a five-point Likert scale. Then the statistical mean was computed using SPSS and then the median was found. Based on the median values, the entire responses were categorized into two groups: high power distance (high PD) and low power distance (low PD), as explained earlier. A similar procedure was adopted to run the path-by-path impact of other variables on an independent variable.

- **Path-by-path moderation test for PD**: The results of the path-by-path of PD are shown here:

  - **Moderation between** SI → BI: PD moderates the relationship between SI → BI (t = 2.418**, p<0.01). At low PD this relationship is *insignificant* (0.004 at p = 0.945) and for high PD this relationship is *significant* (0.271 at p = 0.000).

  - **Moderation between** M → BI: PD moderates the relationship between M and BI (t = 0.254*, p<0.1). At low PD, this relationship is *insignificant* (0.077 at p = 0.254) and for high PD, this relationship is *significant* (0.167 at p = 0.000).
  - **Moderation between** FC → BI: PD moderates the relationship between FC and BI (t = 2.003**, p<0.05). At low PD this relationship is *significant* (0.434 at 0.000) and also for high PD this relationship is *significant* (0.167 at p = 0.024).

- **Path-by-path moderation test for UA**: Path-by-path moderation tests moderate two paths EE → BI and FC → UB.

  - **Moderation between** EE → BI: There is a moderating effect of UA between EE and BI (as t = 2.757***, p<0.01 is observed) such that for high UA (–0.049 at p = 0.610) this relationship is *insignificant*, but for low UA this relationship is *significant* (0.292 at p<0.01). The results show that the UA moderates the relationship between EE and BI.
  - **Moderation between** FC → UB: There is a moderating effect of UA between FC and UB (as t = 3.221***, p<0.01 is observed) such that for high UA (0.294 at p<0.05) this relationship is *significant* but for low UA this relationship is *negatively significant* (–0.230 at p<0.05).
  The results show that the UA moderates the relationship between FC and UB.

- **Path-by-Path Moderation Test for IND**: Path-by-path shows that the IND dimension moderates only one path from PE → BI.

  - **Moderation between** PE → BI: There is a moderating effect of IND between PE → BI (t = 2.264**, p<0.05) such that for high IND (0.144 at p<0.05) this relationship is *significant* and for low PD this relationship is *significant* (0.392 at p<0.01). In other words, at the high IND, the relationship between PE and BI is positively moderated (increased) whereas at low IND this relationship is stronger and almost triple.
  - **Path-by-Path Moderation Test for MAS**: After running the path-by-path test, it was found that MAS has no moderating impact on any of the paths of the model.

7.6.3 Summary of Moderation Tests

Table 7.8 provides a summary of results for the tests performed for moderations at model-level and moderation at the path-by-path level.

Table 7.8 Summary table of moderations (age, experience, and racial groups)

Moderation	Model Level (Moderation)						
	Age	Experience	Awareness	Hofstede's Cultural Dimensions			
				Power Distance	Individualism	Masculinity	Uncertainty Avoidance
	No	Yes	Yes	Yes	No	No	Yes

Paths	Path-by-Path Level (Moderation)					
	Experience	Awareness	Hofstede's Cultural Dimensions			
			Power Distance	Individualism	Masculinity	Uncertainty Avoidance
PE→BI	No	No	Yes	Yes	No	No
EE→BI	No	Yes	Yes	No	No	Yes
FC→BI	Yes	No	No	No	No	No
SI→BI	No	Yes	No	No	No	No
HM→BI	No	Yes	Yes	No	No	No
FC→UB	Yes	No	No	No	No	Yes
H→UB	No	No	No	No	No	No
Bi→UB	No	No	No	No	No	No

'Yes', in the above table means that Path is moderated and 'No' means that paths are not moderated

7.7 Summary

This chapter shows the quantitative analysis of the data obtained from the survey questionnaire. This chapter started with descriptive analysis, data normality tests, reliability tests, and validity tests. This chapter also included EFA and CFA tests for all individual constructs and then the overall model. The statistics of CFA show goodness of fit model. Finally, the structural path model was executed. In this chapter, moderation tests (of age, experience, and cultural dimension) were also executed to the impact of moderators on the relationship between dependent variables and behavioral intention to use LMS technology.

References

Anderson, J., & Gerbing, D. (1988). Structural equation modeling in practice: A review and recommended two-step approach. *Psy- Chol Bull, 103*, 411–423.

Azzalini, A., & Capitanio., A. (1999). Statistical applications of the multivariate skew normal distribution. *Journal of the Royal Statistical Society: Series B (Statistical Methodology), 61*(3), 579–602. http://www.jstor.org/stable/2680724?seq=1#page_scan_tab_contents.

Bagozzi, R. P., Yi, Y. P. L. (1991). Assessing construct validity in organizational research. *Admin Sci Q, 36*(421–58).

Baumgartner, H., & Homburg, C. (1996). Applications of structural equation modeling in marketing and consumer research. *A Review', International Journal of Research in Marketing, 13*, 139–161.

Bentler, P. (1990). Comparative fit indexes in structural models. *Psychological Bulletin, 107*, 238–246.

Blunch, N. J. (2013). *Introduction to structural equation modeling using IBM SPSS statistics and amos.* SAGE Publications.

Byrne, B. M. (2010). *Structural equation modeling with amos: Basic concepts, applications, and programming* (2nd ed.). Taylor and Francis Group.

Cattell, R. B. (1966). The scree test for the number of factors. *Multivariate Behavioral Research, 1*(2), 245–276.

Field, A. (2009). *Discovering statistics using SPSS.* Sage publications.

Fındık, D., & Özkan, S. (2010). *Identifying success factors for WBLMS use by instructors of engineering departments* (pp. 1–7).

Gaskin, J. (2012). *Stats Wiki and Stats tools package.* http://statwiki.kolobkreations.com/.

George, D., & Mallery, P. (2003). *SPSS for windows step by step: A simple guide and reference* (4th ed.). Allyn and Bacon.

Green, S. B., & Salkind. (2005). *Salkind. Using SPSS for windows and macintosh.* Prentice Hall.

Hair, J. F., Black, W. C., Babin, B. J., & Anderson, R. E. (2010). *Multivariate data analysis: A global perspective.* Basım: Pearson Education Inc.

Hussain, M., Khan, M., & Al-Aomar, R. (2016). A framework for supply chain sustainability in service industry with confirmatory factor analysis. *Renewable and Sustainable Energy Reviews, 55*, 1301–1312. https://doi.org/10.1016/j.rser.2015.07.097.

Im, Il Hong, & Seongtae Kang, M. S. (2011). An international comparison of technology adoption testing the UTAUT model. *Information and Management, 48*(1), 1–8. https://doi.org/10.1016/j.im.2010.09.001.

Jöreskog, K. G., & Sörbom, D. (1981). *LISREL V: Analysis of linear structural relations by the method of maximum likelihood.* International Educational Services.

Klem, L. (2000). *Structural equation modeling.*

McCullough, J. & Best, R. (1979). Conjoint measurement: Temporal stability and structural reliability. *Journal of Marketing Research*, 26–31.

Osborne, J. W., & Overbay, A. (2004). The power of outliers (and why researchers should always check for them). Practi. *Practical Assessment, Research and Evaluation, 9*(6), 1–12. http://pareonline.net/getvn.asp?v=9&n=6+.

Pallant, J. (2013). *SPSS survival manual.* McGraw-Hill Education.

Peter, J. P. (1979). Reliability: A review of psychometric basics and recent marketing practices. *Journal of Marketing Research*, 6–17.

Podsakoff, P. M., MacKenzie, S. B., Lee, J. Y., & Podsakoff, N. P. (2003). Common method biases in behavioral research: A critical review of the literature and recommended remedies. *Journal of Applied Psychology, 88*(5), 879–903.

Podsakoff, P. M., & Organ, D. W. (1986). Self-reports in organizational research: Problems and prospects. *Journal of Management, 12*(4), 531–544.

Ritter, G., & Gallegos, M. (1997). Outliers in statistical pattern recognition and an application to automatic chromosome classification. *Pattern Recognition Letters, 18*(6), 525–539. https://doi.org/10.1016/S0167-8655(97)00049-4.

Steiger, J. H. (1990). Structural model evaluation and modification: An interval estimation approach. *Multivariate Behavioral Research, 21*, 309–331.

Straub, D., Boudreau, M.-C., & Gefen, D. (2004). Validation guidelines for IS positivist research. *Communications of the Association for Information Systems, 13,* 380–427.

Tabachnick, B. G., & Fidell, L. S. (2007). *Using multivariate statistics* (5th ed.). Allyn and Bacon. Taylor.

Thompson, B. (2004). *Exploratory and confirmatory factor analysis: Understanding concepts and applications*. American Psychological Association.

Townsend, J., Yeniyurt, S., Deligonul, S., & Cavusgil, S. T. (2004). Exploring the marketing program antecedents of performance in a global company. *Journal of International Marketing, 12*(4), 1–24.

Venkatesh, V., Thong, J., & Xu, X. (2012). Consumer acceptance and use of information technology: Extending the unified theory. *MIS Quarterly, 36*(1), 157–178.

Williams, B., Brown, T., & Onsman, A. (2012). Exploratory factor analysis : A five-step guide for novices. *Journal of Emergency Primary Health Care (JEPHC), 8*(3), 1–13. https://doi.org/10.1080/09585190701763982.

Zou, S., & Cavusgil, S. T. (2002). The GMS: A broad conceptualization of global marketing strategy and its effect on firm performance. *Journal of Marketing, 66*(4), 40–56.

Adoption of LMS: Evidence from the Middle East

8.1 Introduction

This chapter provides empirical evidence regarding instructors' adoption of LMS in higher education institutions (HEIs) in the context of the Middle East in general and Saudi Arabia in particular. The key research questions that have been addressed in this chapter are: (i) To what extent (if any) is behavioral intention (BI) a predictor of use behavior (UB) of LMS technology at Saudi higher educational institutions (SHEIs)?, (ii) To what extent (if any) do independent variables (effort expectancy (EE), performance expectancy (PE), social influence (SI), facilitating condition (FC), hedonic motivation (HM), and habit (H)) impact instructors' behavioral intentions to adopt an LMS at SHEIs?, (iii) Which out of the six independent variables (EE, PE, SI, FC, HM, and H) delivers the most significant contribution to instructors' behavioral intentions to adopt an LMS at SHEIs?, and (iv) To what extent (if any) do moderating variables moderate the relationship between dependent and independent variables? Overall, both direct and indirect critical factors responsible for users' adoption of LMS technology in SHEIs are presented.

8.2 Behavioral Intention and Use Behavior of LMS Technology

Here we address the question: *To what extent (if any) is 'behavioral intention' the predictor of 'use behavior' of LMS technology at HEIs of Saudi Arabia?*

A fundamental concept behind the 'adoption' or 'actual usage' of technology is the 'intention' of an individual (Ajzen, 1991; Taylor & Todd, 1995; Venkatesh & Davis, 2000). The relationship between 'behavioral intention' (BI) and 'use behavior' (UB) has been extensively used in information system (IS) research as a success factor of adoption of the technology (Taylor & Todd, 1995).

UTAUT2 and other models such as UTAUT and TAM hypothesized that the behavioral intention (BI) to use a particular technology is determined by the actual use (UB) of a particular technology. Based on this argument in the literature (see Chap. 2), this study assumes that behavioral intention is a good predictor of user behavior. Accordingly, the research question was designed to validate the relationship between behavioral intention and user behavior. The quantitative results showed that behavioral intention is a significant predictor of the adoption of the LMS. The standardized regression coefficient (SRC) for the path between behavioral intention (BI) and use behavior (UB) was found to be 0.459 (critical ratio = 5.68, p-value < 0.01) showing that BI is a strong predictor of the UB. The finding of this research is also supported by the findings of Alalwan and Williams (2014), Oliveira (2015), and Tosunta, Karada, and Orhan (2015).

In sum, the literature, and the qualitative results show that BI is a strong predictor of UB. This research model explains 58% of the variation in BI and 23.5% of the variation in the user behavior of LMS.

8.3 Influence of Independent Variables on Behavioral Intention (BI)

Here we address the question: *To what extent (if any) do independent variables (PE, EE, SI, FC, HM, and H) impact instructors' behavioral intention to adopt LMS technology at higher educational institutions?*

The purpose of this research question was to examine the influence of independent variables (PE, EE, SI, FC, HM, and H) on the dependent variable (instructors' behavioral intention) to use LMS in their teaching. The influence of independent variables (PE, EE, SI, FC, HM, and H) on the dependent variable (BI) is discussed in the following segment.

© Springer Nature Switzerland AG 2021
R. A. Khan and H. Qudrat-Ullah, *Adoption of LMS in Higher Educational Institutions of the Middle East*,
Advances in Science, Technology & Innovation, https://doi.org/10.1007/978-3-030-50112-9_8

8.3.1 Performance Expectancy (PE)

Performance expectancy (PE) is "*the degree to which an individual believes that using the system will help him or her to attain gains in job performance*" (Venkatesh, Morris, Davis, & Davis, 2003, p. 447). These results reveal that PE is a significant determinant ($p < 0.01$) of BI to adopt the LMS. Of the six independent variables (PE, EE, SI, FC, HM, and H), PE provided the second highest contribution to instructors' behavioral intention to use the LMS. Aside from the above quantitative findings, the literature also supports the concept that PE (i.e. usefulness) has a significant positive correlation with BI. For instance, in a study on computers as a learning tool, Nistor, Lerche, Weinberger, Ceobanu, and Heymann (2014) found that the PE has a positive influence on BI and attitude toward its use (p. E144). In a similar study, using the UTAUT2 model in the teacher's acceptance of LMS, Raman and Don (2013) found that PE has a positive impact on BI. The findings of other researchers (Abdullah & Khanam, 2016; Alalwan & Williams, 2014; Boateng, Mbrokoh, Boateng, & Ansong, 2016; Martins, Oliveira, & Popovič, 2014; Sung, Jeong, Jeong, & Shin, 2015; Wang & Wang, 2010; Wong, Teo, & Russo, 2012; Zhou, Lu, & Wang, 2010) also support this study that PE is strongly correlated to BI. Similarly, in the cultural context of the Middle East, the findings of many researchers such as (Al-Gahtani, Hubona, & Wang, 2007; Al-Somali, Gholami, & Clegg, 2009; Oshlyansky, Park, Cairns, & Thimbleby, 2007) are also similar to this study. Hence, the finding of the above discussion shows that PE (usefulness) of the LMS is highly related to its adoption. This strong determinant predicts that with the increase of benefits, convenience, and advantages (e.g. saving time) of a technology, the acceptance of the technology will also increase. This implies that if the technology is helpful and beneficial, then there are more chances of its adoption. In other words, the more useful LMS is perceived to be, the more the intention to adopt it. Hence the usefulness factor should be considered as an important variable when developing and implementing the LMS at HEIs.

8.3.2 Effort Expectancy (EE)

Effort expectancy (EE) is "the degree of ease associated with the use of the system" (Venkatesh et al., 2003, p. 450). This construct is related to the relationship between instructors' perception of how easy it is to learn and how easy to use LMS, and how their perceptions affect their behavioral intention to use LMS in teaching.

The results of this study showed that effort expectancy (EE) is a significant predictor of BI ($p < 0.01$), which indicates a positive and significant value. This implies that the intention of using technology will increase if users perceive that particular technology is easy to use (Carlsson, Carlsson, Hyvönen, Puhakainen, & Walden, 2006). The finding of this research is consistent with prior research in some cases but inconsistent in other cases. For instance, the finding of this study is not consistent with the findings of Abdullah and Khanam (2016), Carter and Belanger (2004), Kang, Liew, Lim, Jang, and Lee (2015), and Yang (2013). On the other hand, the finding of this research is consistent with (Gawande, 2015; Raman & Don, 2013; Tosunta et al., 2015; Wong et al., 2012; Yun, Han, & Lee, 2013). The results of this research lead to the inference that ease of use of an LMS is an important factor that influences the behavioral intentions of instructors to adopt an LMS. This suggests that instructors having highly positive perceptions of the ease of use of the LMS or the instructors who feel comfortable using LMS would have strong intentions to adopt an LMS system. Hence, if the technology is easy and straightforward to understand, then there are more chances of its adoption.

8.3.3 Social Influence (SI)

Social influence (SI) is based on the supposition that user behavior is influenced by his/her perception of how his/her usage of technology is viewed by other people (Venkatesh, Thong, & Xu, 2016). The results show that SI has a positive and significant link with BI ($p < 0.01$) showing that SI is a strong predictor of BI. Some researchers (such as Anderson, Schwager & Kerns, 2006; George, 2004) found an insignificant or weak relation of SI with BI. However, the finding of this study is supported by some researchers. Similarly, the findings of this study are also supported by the research conducted in non-Western countries by researchers such as (Al-Gahtani et al., 2007; Al-Somali et al., 2009; Tosunta et al., 2015).

This research and the previous research on intentions toward technology have revealed that the influence of society is an important and critical aspect that influences personal beliefs to make decisions about technology adoption (Anderson, Al-Gahtani, & Hubona, 2011). In this context, it is important to consider the role of peers, teams, and groups when implementing LMS. The social influence includes what other people think of the use of technologies as well as the support provided by the top management about the use of an LMS. However, the peer social influence was developed from colleagues or peers who used the LMS. This implies that the influence of other instructors may help in the adoption of LMS. Thus, peer pressure was a key factor that influenced instructors to integrate LMS into their teaching. This shows that if the usage of the LMS is mandatory and enforced by the top management, then everyone will use it. It was also revealed that some instructors use LMS due to the fear of bad evaluation by their students.

8.3.4 Facilitating Conditions (FC)

A facilitating condition (FC) is "the degree to which an individual believes that an organizational and technical infrastructure exists to support the use of the system" (Venkatesh et al., 2003, p. 453). In the quantitative analysis, two links (FC→BI and FC→UB) of the model were examined. The relationship between FC to use behavior (UB) was a non-significant link. This is an unexpected finding and is inconsistent with the findings of Venkatesh, Thong, and Xu (2012) who argued that FC influences both BI and UB. Similarly, many researchers such as (Baptista & Oliveira, 2015; Gawande, 2015; Im, Hong, & Kang, 2011) have found that FC does not influence UB. However, other researchers (Abdullah & Khanam, 2016; Al-Gahtani et al., 2007; Al-Somali et al., 2009; Alalwan & Williams, 2014; Kang et al., 2015; Oshlyansky et al., 2007; Tosunta et al., 2015; Wu, Hsu, & Hwang, 2007) have concluded that the facilitating conditions had positive effects on the usage (UB) of the technology. Hence, there is somewhat of a contradiction in previous studies concerning the relationship between FC and UB. Many researchers (e.g. Bandyopadhyay & Fraccastoro, 2007) even excluded FC from their research study. However, this study found that FC is a positive and significant predictor of BI ($p < 0.05$). Of the six UTAUT2's independent variables (i.e. PE, EE, SI, FC, HM, and H), FC provides the highest (0.272) contribution to instructors' BI to adopt the LMS. The majority of the past studies (AbuShanab, Pearson, & Setterstrom, 2010; Eckhardt, Laumer, & Weitzel, 2009; Kang et al., 2015; San Martin, & Herrero, 2012; Tosunta et al., 2015) support these findings that there is a significant association between FC and BI. Thus, in the quantitative study, facilitating conditions found to have the strongest relationship with BI and act as enablers or motivators of LMS adoption. The technical support is a significant factor in the adoption of new technology (Porter & Graham, 2016). The significant results of facilitating conditions reflect that the existence of technical support and supportive infrastructure is of prime importance to help and support instructors to use LMS in their teaching.

8.3.5 Hedonic Motivation (HM)

Hedonic motivation (HM) is "the fun or pleasure derived from using technology" and "plays an important role in determining technology acceptance and use" (Venkatesh et al., 2012, p. 161). The quantitative results show a strong

and positive relationship between hedonic motivation (HM) and behavioral intention ($p < 0.01$). These results show a strong significant positive relationship between hedonic motivation and behavioral intention to use an LMS. The finding of this study is supported by previous researchers, such as (Alalwan & Williams, 2014; Baptista & Oliveira, 2015; Kang et al., 2015; Raman & Don, 2013; Vinodh & Mathew, 2012; Yang, 2013; Venkatesh et al., 2012). The reason for enjoyment could be due to the usefulness and novel features of the LMS such as chatting, interaction, and instant feedback of LMS, which leads instructors and students to experience pleasure and to consider using this technology to be enjoyable. In sum, it can be concluded that the hedonic motivation has a strong relationship with BI and is one of the core constructs of LMS adoption. This implies that the possibility of LMS adoption will increase among instructors who believe that using LMS is enjoyable, pleasurable, and entertaining.

8.3.6 Habit (H)

Habit is the automaticity of behavior connected with the use of technology over time. The previous use of technology becomes a habit, and habit becomes a strong predictor of future use (Kim & Malhotra, 2005). Previous studies show that an individual's habit is a major predictor of intention to use a technology (Baptista & Oliveira, 2015). In the quantitative analysis, construct H (habit) showed poor values of reliability and convergent validity. This might be because there is no comparison of the functionality of an LMS with a smartphone that is used extensively in our everyday life. This reflects that unlike smartphones, LMS is a teaching tool used only for teaching. Hence, it can be stated that the construct habit might be a valid construct in the context of mobile phones or smart devices but not in the case of LMS. Perhaps, this reason could be the cause of poor reliability and low significance of the habit construct in the quantitative analysis. Hence, this construct should not be part of the LMS adoption model. Therefore, the removal of the 'habit' constructs from the proposed UTAUT2 model would be an appropriate decision in the context of this study.

8.4 The Most Significant Variable(s) for the Adoption of LMS at HEIs

Here we address the question: *Which out of the six independent variables (EE, PE, SI, FC, HM, and H) delivers the most significant contribution to instructors' behavioral intentions to adopt an LMS at SHEIs?*

Of the six independent variables, 'facilitating conditions' (FC) provides the most significant contribution to instructors' behavioral intention. The findings show that technical support and facilitating the environment are given importance by all instructors. This implies that the instructors appear to realize the efforts of management in providing all resources and support. The availability of resources and technical support will encourage the instructors to use the LMS. This, in turn, will improve the instructors' perceptions about using the LMS in their teaching. The second most influential variable is PE. This reflects that instructors realized the LMS to be useful in teaching and learning and have more inclination to use it.

8.5 Demographics and Cultural Dimensions

Here we address the question: *To what extent (if any) do moderating variables, moderate the relationship between the independent and dependent variables?*

A moderator is a quantitative or qualitative variable that influences the strength and direction of the relationship between two other variables (Kripanont, 2007; Lakhal, Khechine, & Pascot, 2013; Schaper, & Pervan, 2007). The moderators of the original UTAUT2 model are age, gender, and experience (Baptista & Oliveira, 2015; Venkatesh et al., 2012). However, for this study, the UTAUT2 model is extended with 'technology awareness' (Hall & Hord, 2011) and 'cultural dimensions' (Bandyopadhyay & Fraccastoro, 2007; Al Gahtani et al., 2007) as moderating variables (Kripanont, 2007; Tibenderana, & Ogao, 2009).

Here we will discuss the findings of age, experience, awareness, and cultural dimensions as the moderators of the UTAUT2 model.

8.5.1 Age as a Moderator of LMS Adoption

In the overall moderation test using SPSS/AMOS, it was found that at the model level, there is no moderating impact of age (insignificant $p = 0.169$) on behavioral intention to adopt LMS. The literature also shows conflicting findings of the age as a moderating variable. Burton-Jones and Hubona (2006) found that age is a significant moderator (Al-Gahtani et al., 2007; Tosunta et al., 2015; Venkatesh et al., 2012) in one technology but insignificant (Hsbollah & Idris, 2009) in another technology. Therefore, there is a need to investigate qualitatively whether age plays an important role in technology adoption (Gibson, Harris, & Colaric, 2008). However, in this study, age is not one of the moderators of BI showing that there is no impact of age on the adoption of LMS.

8.5.2 Experience as a Moderator of LMS Adoption

In an overall moderation test at the model level, it was found that experience moderates (significant $p = 0.01$) the effect of independent variables on the behavioral intention. This shows that there is a difference between two groups of instructors (of low-experience and high-experience) in the adoption of an LMS. The path-by-path test discovered that experience moderates the effect of FC on BI and FC on UB. Experience moderates the effect of *FC on UB* ($t = 2.8***$, $p < 0.01$). It was found that the relationship between FC and UB is significantly moderated both by high experience (0.259 at $p < 0.05$) and low experience (0.187 at $p < 0.01$). Experience moderates the effect of *FC on BI* ($t = 1.661*$, $p < 0.1$). These results show that the relationship between FC and BI is significantly moderated by high experience (0.374 at $p < 0.01$) as well as low experience (0.155 at $p < 0.05$).

The above finding shows that experience affects both paths of FC (i.e. FC to BI and FC to UB). This implied that better technical support and facilitating environments encourage instructors to adopt LMS in their teaching. The finding of this study overlaps with the results of the original model by Venkatesh et al., (2012) and Taylor and Todd (1995). The finding of this study is also supported by the findings of Al-Gahtani et al. (2007), Tosunta et al. (2015), Taylor, and Todd (1995). "It is quite likely that as FC deals with broader infrastructure and support issues, it will always be important to those who value it even if they have significant experience with the target technology" (Venkatesh et al., 2012). Therefore, the experience should be considered to predict the behavioral intention of instructors to use LMS. In sum, the level of experience moderates the overall model (i.e. UTAUT2). It moderates the relationship between FC and BI as well as FC and UB to use the LMS.

8.5.3 Technology Awareness as a Moderator of LMS Adoption

This study uses technology awareness as a moderator of technology adoption as suggested by Faruq and Ahmad (2013), Rehman, Esichaikul, and Kamal (2012), Sun and Fang (2016). In the overall moderation test, the significant p-value ($= 0.099$) shows that awareness has a moderating impact on the behavioral intention at the model level. Furthermore, the path-by-path moderation test showed moderation by three paths: First, the awareness moderates the effect of EE on BI ($t = 2.04**$, $p < 0.05$). It may be noted that at high awareness this relationship (0.035 at $p = 0.703$) is insignificant. However, at low awareness, this relationship is significant (0.272 at $p = 0.000$). This implies that the instructors having low awareness give more importance to ease of use (effort expectancy) of the technology. However, ease of use becomes irrelevant at high-awareness levels. Secondly, the awareness moderates the effect of SI on BI ($t = 2.774***$, $p < 0.01$). At high awareness this relationship ($t = 0.330$ at $p = 0.000$) is significant. At low awareness this relationship is also insignificant ($t = 0.008$ at $p = 0.8977$). Interestingly, high awareness has a stronger impact on the relationship between SI and BI. This implies that with the improved level of awareness, the impact of SI on BI will be higher. Thirdly, the awareness moderates the effect of M on BI ($t = 1.666*$, $p < 0.10$). At high awareness this relationship (0.228 at $p = 0.000$) is significant. However, for low awareness, this relationship is mildly significant (0.088 at $p = 0.109$). This infers that the instructors having high awareness will have a stronger relationship between M and BI than the instructor having low awareness. In other words, high awareness influences the motivational level of the instructors on behavioral intention to use LMS in their teaching. Literature shows that the lack of awareness of the LMS usage exists in higher educational institutions (HEI). Lack of awareness is considered one of the barriers to technology adoption (Shannak, 2013). The results also revealed that among the instructors who are even aware of the LMS, a large number of the instructors are the *light users* of the LMS. Therefore, it can be assumed that the instructors use LMS, they used its very basic functions, or they use the LMS only for posting the grades, which is mandatory from higher management. This shows that although there is awareness of the LMS, it is of a low level. These results demonstrate that the instructors use the LMS but they are utilizing a limited number of features.

8.5.4 Cultural Dimensions as the Moderator of LMS Adoption

Here we address the question: *To what extent (if any) does cultural dimensions (such as PD, IND, MASC, and UA), moderate the relationship between independent variables (PE, EE, SI, FC, HM, and H) and 'behavioral intention' to use the LMS?*

In this study, Hofstede's cultural dimensions (i.e. power distance, individualism/collectivism, masculinity, and uncertainty avoidance) were used as the moderators (Al-Gahtani et al., 2007; Baptista & Oliveira, 2015) of LMS adoption. The following section will discuss the impact of four cultural dimensions on instructors' behavioral intuition to adopt an LMS.

(a) **Power Distance as Moderating Variable**: It was found that the presence of power distance (PD) in a society moderates the instructor's behavioral intention to adopt an LMS. Furthermore, the path-by-path test exhibited that PD moderates three paths (i.e. SI→BI, M→BI, and FC→BI). The finding of this study is similar to researchers such as (Al-Gahtani et al., 2007; Baptista & Oliveira, 2015; Im, Il Hong, & Seongtae Kang, 2011), but other researchers such as (Nistor et al., 2014; Yoon, 2009), found that PD does not behave as a moderator of behavioral intention. It is evident from the results that this society is lacking in the adoption of an LMS. The finding that shows Arabs' society possesses a high PD value in the educational environment. As shown in Fig. 4.2, the PD of Saudi Arabia is high (95) as compared to the UK (35). This implies that in a high PD society the subordinates are told (ordered) to do their jobs and the boss is an autocrat (Al-Gahtani et al., 2007). This shows that a society having a high PD gives more importance to social norms, values, and traditions, which in turn affect the adoption of an LMS. Hence, the results of quantitative results, supported by the previous literature, prove that the power distance exists in the society and it moderates the behavioral intention to use new technology. Thus, it can be established that PD is present in the higher educational institutions of Arab culture that may lead to the lower adoption of the LMS. The existence of the PD is an indication of the low adoption of the LMS.

(b) **Individualism/collectivism (IND) as Moderator**: At the model level, the overall moderation test showed that there is no moderating impact of individualism (IND). The finding of this study is inconsistent with Nistor et al. (2014) but consistent with Yoon (2009), who found that IND has no moderating impact between independent and dependent variables. Hofstede (1980) stated that in greater individualistic societies, an individual is least concerned with the beliefs of other people and, hence, he is less worried to follow any specific behavior of the society. As shown in Fig. 4.2, the IND value for Saudi Arabia is 25, whereas the IND value of the UK is 89. The low value of the individualism index for KSA refers to a culture where collective opinions are more important. In other words, the collective opinions of a group of people have a significant impact on individual behavioral intentions, showing that there is a positive correlation between subjective norm and behavioral influence (Al-Gahtani et al., 2007). The individual is more worried about the opinions of others and hence follows the behavior that is more important to the group. In this kind of culture, loyalty is more dominant, and it overrides most other social rules and regulations. In collectivist societies, the relationships between employers and employees are perceived to be familial in moral terms. In such societies, technology is adopted based primarily on social values. The literature revealed that in the cultures with a greater IDV index, it is easier to adopt new technologies (Van Everdingen & Waarts, 2003) whereas the lack of individualism in any society causes low adoptions of the technology.

(c) **Femininity/Masculinity (MASC) as Moderator**: At the model level, the moderation test shows that there is no moderating effect of a femininity/masculinity dimension. The path-by-path moderation test also revealed that the masculinity/femininity dimension does not moderate any path of the model. Surprisingly, this study is not consistent with other studies. For instance, Yoon (2009) found that masculinity/femininity has a moderating effect on perceived usefulness, ease of use, and behavioral intention. Similarly, Im et al. (2011), Sun and Zhang (2006), and Nistor et al. (2014) discovered that there is an influence of MASC on behavioral intention. Also, as shown in Fig. 4.2 that the score of MASC for KSA is 60, whereas it is 66 for the UK (Hofstede, 2010); meaning that the MASC value for Saudi Arabia is inclined toward the high side. In a masculine culture, people are more goal-oriented. The individuals in such cultures are more concerned with sticking with the absolute truth and are considered as normative in their thinking. A MASC society is known for recognition for improvement, competition, more task-oriented, achievement, and success of the people; and these characteristics of culture lead to the adoption of new technology (Van Everdingen, & Waarts, 2003; Hasan & Ditsa, 1999; Hofstede, 2011).

(d) **Uncertainty avoidance (UA) as moderating variable**: The quantitative study indicated the existence of UA that moderates the adoption of the LMS. The overall moderation test shows that the presence of the UA dimension moderates 'behavioral intention' of the users to use LMS. The findings of this research study are consistent with the other researchers. For instance, Yoon (2009) found that a high uncertain culture may decrease the intention of people for online shopping. Similarly, Nistor et al. (2014) discovered that uncertainty avoidance influences the BI negatively. The path-by-path moderation tests indicated the moderation of two paths. First, the presence of UA influences the relationship between EE and BI. Secondly, the presence of UA influences the relationship between FC and UB.

This finding is supported by Nistor et al. (2014) and Im, Kim, and Han (2008). This implies that the FC deals with the facilitation and support issues and is given importance by users. Hence, if more facilities support the use of an LMS, then instructors would be more likely to use the LMS. Hence, the existence of uncertainty in a society is the indication of low adoptions of the technology (Van Everdingen & Waarts, 2003; Hasan & Ditsa, 1999). As shown in Fig. 4.2, the score of uncertainty avoidance for KSA is 80 (high), whereas it is 35 (low) for the UK. A high UA culture reflects a structured and a rule-oriented society having many rules, regulations, and controls. The countries demonstrating such a high value of UA retain rigid codes of behavior and belief. A culture with a greater UA index will have resistance to adopting any new technology because the society will not able to take the risks of trying new technologies (Im et al., 2011) and will only accept innovations that have already been used by others. Consequently, it will not be easy to adopt new technologies in a culture with greater values of UA (Van Everdingen & Waarts, 2003; Hasan & Ditsa, 1999). Hence, the presence of high UA in the HEIs of Arab societies could cause the lower adoption of LMS technology because the innovations and adoptions of technology are resisted in such societies.

Thus, the results of this study reveal that the Arab community appears to have more 'power distance' and less 'uncertainty avoidance'. This might be one of the reasons for less adoption of LMS.

8.6 Summary of the Empirical Evidence

We developed several research questions to explain and assess the relationships between independent and dependent variables of the UTAUT2 model. The key findings are summarized in Table 8.1.

8.7 UTAUT2 Tested Model

Based on the above discussion, the following tested model (Fig. 8.1) is proposed for the adoption of LMS in higher educational institutions of the Middle East region:

Concerning the adoption of LMS, it appears that the variables PE, EE, SI, FC, and HM directly influence the

Table 8.1 Summary of the findings

RQ	Statement	Findings
1	To what extent (if any) is 'behavioral intention' the predictor of 'use behavior' of LMS technology at HEIs of Saudi Arabia?	The quantitative findings support that behavioral intention is a strong predictor of the actual use of the LMS
2	To what extent (if any) do independent variables (PE, EE, SI, FC, HM, and H) impact instructors' behavioral intention to adopt LMS technology at HEIs?	The constructs PE, EE, SI, FC, and HM were found to be the significant predictors of the behavioral intention, except the construct H
3	This research question was framed to explore which out of the six independent variables (EE, PE, FC, SI, HM, and H) delivers the most significant contribution to instructors' behavioral intention to use an LMS	Of the six UTAUT2's independent variables, FC provides the highest significant contribution to the instructor's BI to use the LMS, while PE provides the second-highest significant contribution to the instructor's BI to use the LMS
4	To what extent (if any) do moderating variables moderate the relationship between the independent and dependent variables?	**Age as a Moderator of LMS adoption**: Age was not found to be the moderator in quantitative analysis **Experience as a moderator of LMS adoption**: Experience was found to be a moderator on the overall model and it moderates the effect of FC to BI and FC to UB **Technology awareness as a moderator of LMS adoption**: TA was found to be a moderator on the overall model and moderates the effect of EE on BI, SI on BI, and M on BI **Power distance as a moderator of LMS adoption**: In the quantitative strand, the power distance was found to be a moderator on the overall model, and it moderates the effect of SI to BI, M to BI, and FC to BI **Uncertainty avoidance as a moderator of LMS adoption:** In the quantitative strand, it was found that UA moderates the model. It moderates the effect of EE on BI, and FC on BI **Individualism as a moderator of LMS adoption**: In the quantitative strand, no moderating impact was found on the model level **Masculinity as a moderator of LMS adoption**: In the quantitative study, no moderating impact was observed on the model level or path level

behavioral intention except for the variable habit (H). Of six variables (i.e. PE, EE, FC, SI, HM, and H), FC provides the most significant contribution to instructors' behavioral intention followed by PE and HM. Thus, while considering the adoption of LMS, FC, PE, and HM are the critical factors and form the basis of any successful adoption of LMS endeavor in the Middle Eastern region.

8.8 Summary

In this chapter, we addressed many research questions about the adoption of LMS technology. We analyzed both direct and indirect factors responsible for adoption LMS in higher educational institutions. Then we presented our tested model,

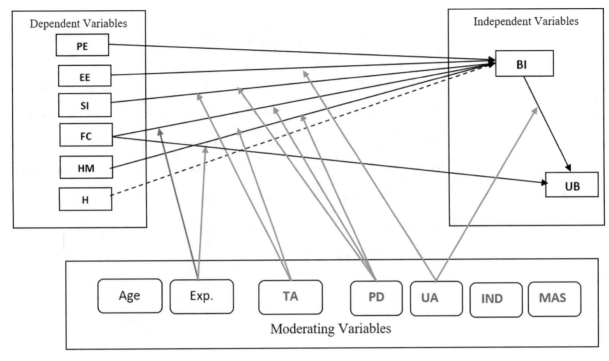

Fig. 8.1 Tested model

the UTAUT2, in the couture context of the Arab community. We found that almost all constructs of the UTAUT2 model (such as PE, EE, FC, SI, and HM) play a significant role in the adoption of LMS technology except the construct 'habit'. The UTAUT2 model was also tested with the inclusion of 'technology awareness' and Hofstede's cultural dimensions as the moderating variables. The quantitative results revealed that experience, technological awareness, and cultural dimensions moderate the relationship between dependent and independent variables of the UTAUT2 model.

References

Abdullah, M., & Khanam, L. (2016). The influence of website quality on m-banking services adoption in Bangladesh : Applying the UTAUT2 model using PLS. *International Conference on Electrical, Electronics, and Optimization Techniques (ICEEOT)*, 1–19.

AbuShanab, E., Pearson, J. M., & Setterstrom, A. (2010). Internet banking and customers' acceptance in Jordan: The unified model's perspective. *Communications of the Association for Information Systems, 26*(1), 493–524.

Ajzen, I. (1991). The theory of planned behavior. *Organizational Behavior and Human Decision Processes, 50*(2), 179–211.

Al-Gahtani, S. S., Hubona, G. S., & Wang, J. (2007). Information technology (IT) in Saudi Arabia: Culture and the acceptance and use of IT. *Information & Management, 44*(8), 681–691. https://doi.org/10.1016/j.im.2007.09.002.

Al-Somali, S. A., Gholami, R., & Clegg, B. (2009). An investigation into the acceptance of online banking in Saudi Arabia. *Technovation, 29*(2), 130–141. https://doi.org/10.1016/j.technovation.2008.07.004.

Alalwan, A., & Williams, M. (2014). Examining factors affecting customer intention and adoption of internet banking in Jordan. *UK Academy for Information Systems Conference, 3.*

Anderson, C. S., Al-Gahtani, S., & Hubona, G. (2011). The value of TAM antecedents in global IS development and research. *Journal of Organizational and End User Computing, 23*(1), 18–37. https://doi.org/10.4018/joeuc.2011010102.

Anderson, J. E., Schwager, P. H., & Kerns, R. L. (2006). The drivers for acceptance of tablet PCs by faculty in a college of business. *Journal of Information Systems Education, 17*(4), 429–440.

Bandyopadhyay, K., & Fraccastoro, K. A. (2007). The effect of culture on user acceptance of information technology. *Communications of AIS, 19*, 522–543.

Baptista, G., & Oliveira, T. (2015). Understanding mobile banking: The unified theory of acceptance and use of technology combined with cultural moderators. *Computers in Human Behavior, 50*, 418–430. https://doi.org/10.1016/j.chb.2015.04.024.

Boateng, R., Mbrokoh, A. S., Boateng, L., & Ansong, P. K. S. E. A. (2016). Determinants of e-learning adoption among students of developing countries. *International Journal of Information and Learning Technology, 33*(4), 248–262. https://doi.org/10.1108/IJILT-02-2016-0008.

Burton-Jones, A., & Hubona, G. S. (2006). The mediation of external variables in the technology acceptance model. *Information & Management, 43*(6), 706–717. https://doi.org/10.1016/j.im.2006.03.007.

Carlsson, C., Carlsson, J., Hyvönen, K., Puhakainen, J., & Walden, P. (2006). Adoption of mobile devices/services—searching for answers with the UTAUT. *39th Hawaii International Conference on System Sciences*, Hawaii, 00(C), 1–10.

Carter, L., & Belanger, F. (2004). Citizen adoption of electronic government initiatives. *37th Annual Hawaii International Conference on System Sciences, 2004. Proceedings of The, 00*(C), 10 pp. https://doi.org/10.1109/HICSS.2004.1265306.

Eckhardt, A., Laumer, S., & Weitzel, T. (2009). Who influences whom? Analyzing workplace referents' social influence on IT adoption and non-adoption. *Journal of Information Technology, 24*(1), 11–24.

Faruq, M. A., & Ahmad, H. B. (2013). The moderating effect of technology awareness on the relationship between UTAUT constructs and behavioural intention to use technology: A conceptual paper. *Australian Journal of Business and Management Research, 3*(2), 14–23. https://doi.org/10.1016/j.im.2013.09.002.

Gawande, V. (2015). Development of blended learning model based on the perceptions of students at higher education institutes in Oman. *International Journal of Computer Applications, 114*(1), 38–45. http://www.ijcaonline.org/archives/volume114/number1/19946-1747.

George, J. F. (2004). The theory of planned behavior and Internet purchasing. *Internet Research, 14*(3), 198–212. https://doi.org/10.1108/10662240410542634.

Gibson, S., Harris, M., & Colaric, S. (2008). Technology acceptance in an academic context: Faculty acceptance of online education. *Journal of Education for Business, 83*(6), 355–359.

Hall, G. E., & Hord, S. M. (2011). Implementing change: Patterns, principles, and potholes. (3rd ed.,). Pearson Education.

Hofstede, G. (1980). Culture's consequences: International differences in work-related values. Sage Publications.

Hofstede, G. J. (2010). *Cultures and Organizations- software of the mind: Intercultural Cooperation and Its Importance for Survival* (Revised an). McGraw Hill.

Hofstede, G. (2011). Dimensionalizing cultures: The hofstede model in context. *Online Readings in Psychology and Culture, 2*(1), 1–26. https://doi.org/http://dx.doi.org/10.9707/2307-0919.1014.

Hsbollah, M. H., & Idris, K. M. (2009). E-learning adoption: the role of relative advantages, trialability and academic specialisation. *Campus-Wide Information Systems, 26*(1), 54–70. https://doi.org/10.1108/10650740910921564.

Im, I., Kim, Y., & Han, H. J. (2008). The effects of perceived risk and technology type on users' acceptance of technologies. *Information & Management, 45*, 1–9.

Im, Il Hong, & Seongtae Kang, M. S. (2011). An international comparison of technology adoption Testing the UTAUT model. *Information & Management, 48*(1), 1–8. https://doi.org/10.1016/j.im.2010.09.001.

Kang, M., Liew, B. T., Lim, H., Jang, J., & Lee, S. (2015). Investigating the determinants of mobile learning acceptance in Korea Using UTAUT2. In *Emerging issues in smart learning* (pp. 209–216). https://doi.org/10.1007/978-3-662-44188-6.

Kim, S. S., & Malhotra, N. K. (2005). A Longitudinal model of continued is use: An integrative view of four mechanisms underlying postadoption phenomena. *Management Science, 51*(5), 741–755. https://doi.org/10.1287/mnsc.1040.0326.

Kripanont, N. (2007). Using technology acceptance model of Internet usage by academics within Thai business schools [Victoria University]. http://wallaby.vu.edu.au/adt-VVUT/public/adtVUT20070911.152902/index.html.

Lakhal, S., Khechine, H., & Pascot, D. (2013). Student behavioural intentions to use desktop video conferencing in a distance course: Integration of autonomy to the UTAUT model. *Journal of Computing in Higher Education, 25*(2), 93–121. https://doi.org/10.1007/s12528-013-9069-3.

Martins, C., Oliveira, T. & Popovič, A. (2014). Understanding the internet banking adoption: A unified theory of acceptance and use of technology and perceived risk application. *International Journal of Information Management, 34*(1), 1–13.

Nistor, N., Lerche, T., Weinberger, A., Ceobanu, C., & Heymann, O. (2014). Towards the integration of culture into the Unified Theory of Acceptance and Use of Technology. *British Journal of Educational Technology, 45*(1), 36–55. https://doi.org/10.1111/j.1467-8535.2012.01383.x.

Oshlyansky, L., Park, S., Cairns, P., & Thimbleby, H. (2007). Validating the unified theory of acceptance and use of technology (UTAUT) tool cross-culturally. *Proceedings of the 21st British HCI Group Annual Conference on People and Computers, 2* (September).

Porter, W. W., & Graham, C. R. (2016). Institutional drivers and barriers to faculty adoption of blended learning in higher education. *British Journal of Educational Technology, 47*(4), 748–762. https://doi.org/10.1111/bjet.12269.

Raman, A., & Don, Y. (2013). Preservice teachers' acceptance of learning management software: An application of the UTAUT2 model. *International Education Studies, 6*(7). https://doi.org/10.5539/ies.v6n7p157.

Rehman, M., Esichaikul, V., & Kamal, M. (2012). Factors influencing e-government adoption in Pakistan. *Transforming Government: People, Process and Policy, 6*(3), 258–282. https://doi.org/10.1108/17506161211251263.

San Martin, H., & Herrero, A. (2012). Influence of the user's psychological factors on the online purchase intention in rural tourism: Integrating innovativeness to the UTAUT framework. *Tourism Management, 33*(2), 341–350.

Schaper, L. K., & Pervan, G. P. (2007). ICT and OTs: A model of information and communication technology acceptance and utilisation by occupational therapists. *International Journal of Medical Informatics, 76*, S212–S221.

Shannak, R. O. (2013). The difficulties and possibilities of E-Government: The case of Jordan. *Journal of Management Research, 5*(2), 189–204. https://doi.org/10.5296/jmr.v5i2.2560.

Sun, H., & Fang, Y. (2016). Choosing a fit technology: understanding mindfulness in technology adoption and continuance. *Journal of the Association for Information Systems, 17*(6), 377.

Sun, H., & Zhang, P. (2006). The role of moderating factors in user technology acceptance. *International Journal of Human Computer Studies, 64*(2), 53–78. https://doi.org/10.1016/j.ijhcs.2005.04.013.

Sung, H., Jeong, D., Jeong, Y., & Shin, J. (2015). The effects of self-efficacy and social influence on behavioral intention in mobile learning service. *Education, 103,* 27–30.

Taylor, S., & Todd, P. (1995). Understanding information technology usage: A test of competing models. *Information Systems Research, 6*(4), 144–176.

Tibenderana, P. K., & Ogao, P. (2009). Information technologies acceptance and use among universities in Uganda: A model for hybrid library services end-users. *International Journal of Computing and ICT Research, 1*(1), 60–75. http://www.ijcir.org/.

Tosunta, B., Karada, E., & Orhan, S. (2015). The factors affecting acceptance and use of interactive whiteboard within the scope of FATIH project: A structural equation model based on the Unified Theory of acceptance and use of technology. *Computers & Education, 81,* 169–178. https://doi.org/10.1016/j.compedu.2014.10.009.

Van Everdingen, Y. M., & Waarts, E. (2003). The effect of national culture on the adoption of in- novations. *Marketing Letters, 14*(3), 217–232. https://doi.org/10.1023/A:1027452919403.

Venkatesh, V., & Davis, F. D. (2000). A theoretical extension of the technology acceptance model: Four longitudinal field studies. *Management Science, 46*(2), 186–204.

Venkatesh, V, Thong, J., & Xu, X. (2012). Consumer acceptance and user of information technology: Extending the unified theory of acceptance and use of technology. *MIS Quarterly, 36*(1), 157–178. Retrieved from http://ezproxy.library.capella.edu/login?url=. http://search.ebscohost.com/login.aspx?direct=true&db=iih&AN=71154941&site=ehost-live&scope=site.

Venkatesh, V., Morris, M. G., Davis, G. B., & Davis, F. D. (2003). *User Acceptance of Information Technology: Toward A Unified View. 27*(3), 425–478.

Venkatesh, V., Thong, J. Y. L., & Xu, X. (2016). Unified theory of acceptance and use of technology: A synthesis and the road ahead. *Journal of Association for Information Systems, 17*(5), 328–376.

Vinodh, K., & Mathew, S. K. (2012). Web personalization in technology acceptance. In Intelligent Human Computer Interaction (IHCI). *Th Int. Conf. Intell. Hum. Comput. Interact IEEE,* 1–6.

Wang, H. Y, & Wang, S. W. (2010). User acceptance of mobile internet based on the unified theory of … *Social Behavior and Personality,* 415–426.

Wong, K.-T., Teo, T., & Russo, S. (2012). Interactive whiteboard acceptance: Applicability of the UTAUT model to student teachers. *The Asia-Pacific Education Researcher, 22*(1), 1–10. https://doi.org/10.1007/s40299-012-0001-9.

Wu, H., Hsu, Y., & Hwang, F. (2007). Factors affecting teachers adoption of technology. September 2006, 63–85.

Yang, S. (2013). Understanding undergraduate students' Adoption of mobile learning model: A perspective of the extended UTAUT2. *Journal of Convergence Information Technology, 8*(10), 969.

Yoon, C. (2009). The effects of national culture values on consumer acceptance of e-commerce: Online shoppers in China. *Information & Management, 46*(5), 294–301. https://doi.org/10.1016/j.im.2009.06.001.

Yun, H., Han, D., & Lee, C. C. (2013). Understanding the use of location-based service applications: Do privacy concerns matter? *Journal of Electronic Commerce Research, 14*(3), 215–231.

Zhou, T., Lu, Y., & Wang, B. (2010). Integrating TTF and UTAUT to explain mobile banking user adoption. *Computers in Human Behavior, 26*(4), 760–767. https://doi.org/http://dx.doi.org/10.1016/j.chb.2010.01.013.

9.1 Introduction

This chapter presents a summary of this book project. This provides an overview of the key elements of research that was undertaken in this project. The contributions and insights toward LMS adoption, implications for theory and practice regarding the use and adoption of the learning management system, and the limitations of this study are discussed. Finally, the recommendations for future research are made.

9.2 Background of Research

This study developed a research model based on an extensive literature review of existing theories and models on technology adoption in the cultural context of higher educational institutions. Numerous technology adoption models such as technology acceptance model (TAM) by Davis (1989), diffusion of innovation (DOI) by Rogers (1995), theory of reasoned action (TRA) by Fishbein and Ajzen (1975), and unified theory of acceptance and use of technology (UTAUT) by Venkatesh, Morris, Davis, and Davis (2003) were reviewed. A modified UTAUT model (i.e. UTAUT2) was adapted for this research. Based on the literature review, this research extended the UTAUT2 model with Hofstede's (Hofstede, 1980) 'cultural dimensions', and 'technology awareness' as the moderators of the model. In this quantitative study, the data were collected via a survey questionnaire.

The sample for the quantitative study belongs to a target population of 2005 instructors of higher educational institutions. The quantitative data utilized structural equation modeling (SEM) for 310 valid users of the learning management system (LMS) to assess the adequacy of the measurement model. The result of the analysis revealed that all constructs of the UTAUT2 play a significant role in the adoption of LMS technology except the construct 'habit'. The model was also tested with the inclusion of 'technology awareness' and Hofstede's cultural dimensions as the moderating variables. In moderating variables testing, it was found that 'experience', 'technological awareness', and 'cultural dimensions' play a significant impact on the relationship between dependent and independent variables of the model.

The following sections of the chapter provide the contributions, implications, areas for future research, and the recommendations.

9.3 Advancing the Theory and Practice of LMS Adoption

The research study in this book makes the following contributions:

- **Advancing the knowledge about LMS adoption and usage**: The original UTAUT2 model by Venkatesh et al. (2012) was developed and tested in non-Western countries. The literature is lacking in the implementation of this model in the adoption of LMS for instructors in the cultural context of higher educational institutions. Thus, this study contributes significantly to the body of knowledge through the inclusion of a study on UTAUT2 in the cultural context of higher educational institutions. This study also identifies the significant factors such as 'performance expectancy' (PE), 'effect expectancy' (EE), 'social influence' (SI), 'facilitating conditions' (FC), habit (H), and 'hedonic motivation' (HM) that influence instructors' behavioral intentions to adopt LMS. This research has extended the model into the following three dimensions: The first extension is the 'technological' extension. In this study, 'LMS technology' was considered as a technological extension. The second extension is the new 'user population' extension. The focus of this study was the 'instructors' of higher educational institutions. The third extension of the model is the inclusion of 'additional variables'. This research extended the

© Springer Nature Switzerland AG 2021
R. A. Khan and H. Qudrat-Ullah, *Adoption of LMS in Higher Educational Institutions of the Middle East*,
Advances in Science, Technology & Innovation, https://doi.org/10.1007/978-3-030-50112-9_9

UTAUT2 model with moderating variables. Technological 'awareness' and Hofstede's (Hofstede, 1980) 'cultural dimensions' were used as moderating variables of the model. Although variables of the model are adapted from the literature and the original UTAUT2 model, the proposed model is novel because this is the first study of its kind which extended the UTAUT2 model into three unique dimensions in exploring instructors' intention to use LMS in the cultural context of higher educational institutions. In comparison to other models such as DOI, UTAUT, and TAM, the proposed model is the most comprehensive one.

- **Implications for theory and practice of LMS adoption**: The output of this research has practical and theoretical implications. It provides a base for educational managers to consider how to implement LMS systems. This is particularly relevant for areas of academia with low levels of LMS use. Also, the work contributes to a theoretical level to the use and extension of adoption theories relevant to understanding technology adoption within corporate educational environments.

- **Research implications**: The extension of the model into new dimensions in the UTAUT2 model suggests new avenues for the researchers in the area of technology adoption models. The findings of this research will enhance the understanding of the practicability of the UTAUT2 model across multiple settings and cultures. It is believed that the model could be replicated in other regions of the world. This study offers the administrators of higher educational institutions with the facility to recognize the variables that influence the instructors' intention to adopt LMS and to incorporate these influential variables into the planning, investment, and implementation for effective adoption of LMS. This study adapted the original instrument developed by Venkatesh et al. (2012) and Hofstede's cultural dimensions with some modifications suitable for educational and cultural settings. Thus, this study provided new instruments to assess the predictors of LMS in the cultural context of higher educational institutions. It is expected that the methodology used for this study can be used as a guide for new researchers to aid in assessment of the predictors of technology adoption.

- **Practical Implications**: The findings of this study lead to the conclusion that 'ease of use' of an LMS is an important factor that influences the behavioral intentions of instructors to adopt an LMS. If the technology is easy and straightforward to understand, then there is more chance of its adoption. This suggests that instructors having highly positive perceptions of 'ease of use' of the LMS would have strong intentions to adopt an LMS system. The finding indicated that 'usefulness' emerged

as one of the most influential variables influencing instructors' behavioral intention to adopt LMS at higher educational institutions. This shows that instructors will use LMS if they find it useful in their teaching process. This implies that with the increase of benefits, convenience, and advantages (e.g. time-saving) of a technology, the more the likelihood of the acceptance of the technology will also increase. There is consequently a link between the perceived utility of the LMS and the increase in the intention to adopt it. Therefore, the LMS designers should design the LMS so that the instructors should feel that LMS is useful and comfortable for their teaching. A clear vision of the usefulness and positive features of the LMS conveyed to the instructors and students would motivate them to use LMS in their teaching and learning. By using these results, the administrators of higher educational institutions may launch technology awareness programs, seminars, workshops, to convince instructors how easy to use and useful this new technology is for them. This thereby encourages an increase in usage of LMS in their teaching and learning processes. Similarly, LMS designers can use the findings to reassess and redesign the interface structure of the LMS. To promote their product, LMS designers may need to ensure that their systems are simple, easy-to-use, user-friendly, and compatible with all teaching and learning environments.

This research confirmed that *facilitating conditions (FC)* and availability of resources are the significant predictors influencing the behavioral intention to adopt the LMS. The lack of required resources and facilitating conditions for instructors act as a barrier and demotivate instructors from using LMS in HEIs (Al-Gahtani et al. 2007). The management should consider providing all necessary technical support and facilitating conditions such as necessary hardware, software, and also offer training workshops on LMS. This will enhance their level of comfort in using LMS. The findings and recommendations of this study would allow the management of higher educational institutions to re-evaluate their existing practices and support systems for LMS. This, in turn, will improve the instructors' perceptions about using the LMS in their teaching.

In this study, *technology awareness (TA)* is introduced as a new moderator and was found to be a significant moderator of the model. The study showed that technology awareness plays an important role in the adoption of new technology. This suggests the need to raise awareness of modern technological developments and would be beneficial to all stakeholders (instructors, administrators, and students) in promoting the awareness and benefits of the LMS in the teaching and learning process. The technology awareness campaign by higher management will help the instructors to be aware of the benefits of the LMS and hence, will help improve the adoption of the LMS.

The findings show a strong and positive correlation between *hedonic motivation (HM)* and behavioral intention, indicating that the chances of LMS adoption will be increased among instructors who believe that using an LMS is more enjoyable, pleasurable, and entertaining. This implies that instructors achieve an acceptable level of intrinsic motivation while using LMS in their teaching. Hence, it is suggested that instructors should provide some components of fun in their teaching activities such as games, online quizzes, and videos relevant to the course content to let students feel entertained and playful while using the LMS. Also, the designers and developers of the LMS should design the interface and features of LMS so that the instructors feel LMS is a user-friendly, comfortable, and entertaining tool rather than a burden.

This study shows that *culture* is an important factor that influences the technology adoption. To bring increased acceptance of the new technology, the technology should be designed to match with the cultural values of the society. The designers and developers of LMS should consider the cultural values and dimensions (i.e. PD, UA, IND, and MAS) of a community while developing an LMS for a community. The findings of this study could also serve as a guide for other institutions to understand the impact of cultural variables on the LMS adoption from the instructors' point of view. The findings of this research and the literature study show that Arab culture has a high power distance. This implies that the managers have a lower tendency to exchange and share information. The information gap acts as a barrier to the adoption of LMS among instructors. The management of the higher educational institutions should promote information sharing environment to reduce the information gap. The findings indicate that most instructors are uncertain about the safety and security of the students' data stored on the LMS. The concerns about seamless availability, privacy, and security of the students' data stored on the LMS continues to be one of the fears of the instructors in the adoption of the LMS. Hence, the management of the higher educational institutions should build the trust of instructors regarding technology by investing in technology improvement, data privacy, and security. LMS providers should incorporate modern technological advances (e.g. cloud) to secure the data on the LMS to make instructors more comfortable and to have more confidence in the technology. This trust and confidence would convey a positive impact on the increasing participation of the instructors to use the LMS.

9.4 Future Research Directions

The following suggestions are made for future research:

- This study provides an empirical base to develop an extended UTAUT2 model by incorporating additional variables in a multi-cultural context and provides a guide for further study. The proposed model could be further tested in the future with a larger sample of instructors in all higher educational institutions of the Middle East. Similarly, the methodology used for this model can be replicated in future research in higher educational institutions of other regions of GCC countries.

- This study involves a one-time data collection (cross-sectional). In the future, longitudinal studies could be conducted for better observing and evaluating the changes in the variables influencing the instructors' behavioral intention to adopt LMS in their teaching.

- This study assumes that behavioral intention (BI) is a good predictor of use behavior (UB). In this regard, in the future, there is a need to reassess whether the 'behavioral intention' predicts 'use behavior'. Furthermore, in the UTAUT2 model, use behavior (UB) is dependent upon behavioral intention (BI) and is measured indirectly through BI. In the future, research can be conducted on other models in which the adoption (i.e. UB) is directly measured rather than through BI.

- This research focused only on Hofstede's four cultural dimensions. Future research could expand the research to examine the role of long term/short term and indulgence/restraint dimensions. Also, this study has dealt with the impact of culture on the LMS adoption at an individual level, but in the future, to achieve more focused results, the research could be focused on a national as well as on an individual level. Another limitation could lie in the utilization of, the 30-year-old, Hofstede's cultural dimensions which was introduced in 1980. In 30 years, many changes could have occurred in the cultural values of the nations. Therefore, a more recent study on Hofstede's cultural dimensions could provide a better comparison with this research.

- Since the sample is limited to only male instructors, future research could be extended to male and female populations.

- This research has been conducted only on the instructors. Future research could be extended by incorporating the perceptions of a student, as well.

- The original UTAUT2 model shows that FC is a major predictor of both BI and UB. But this study shows an insignificant relationship between FC and UB. This finding contradicts the finding of the original model and needs to be retested in the future to determine the reason for it not being significant.

- The focus of this study was limited to the investigation of the original constructs of the UTAUT2 model and moderating variables (age, experience, technology awareness, and Hofstede's cultural dimensions) as influential variables of LMS adoption. Future studies could explore and analyze all the possible variables (such as system quality, students' feedback, organizational factors, instructors' attitude, and curriculum) which may influence the adoption of LMS. This would give a more complete picture of all of the factors that influence the adoption of LMS.

9.5 Macro-Level and Micro-Level Recommendations

Based on findings from this research study, the recommendations are made and classified into two (micro and macro) levels for possible improvement of LMS adoption at higher educational institutions.

The first level (macro-level) includes recommendations for the management of higher educational institutions, LMS designers and the Ministry of Education (MoE) while the second level (micro-level) recommendations include the recommendations for instructors and researchers.

9.5.1 Macro-Level Recommendations

1. **Providing Required Technical Support and Facilitating Conditions**: This research showed that facilitating conditions and availability of resources are the most significant predictors of LMS adoption. Therefore, management should consider providing all necessary technical support and facilitating conditions, such as offering training workshops on LMS. The management should facilitate the quick access of instructors to the resources needed for the use of LMS. Furthermore, because of the rate at which technological development and growth are taking place, prospective institutions may find it important to have round the clock continuous technical support. For this purpose, training can be organized in the institutions, call centers can be established to get immediate solutions to problems or a continuous consultancy may be offered to the instructors (Tosunta, Karada, & Orhan, 2015). Also, walk-in professional development programs on instructional technology, online courses, and technology-related topics could be organized for all instructors of higher

educational institutions. The availability of technical support and workshops improve their level of comfort in using LMS. Hence, the policymakers should consider well-organized efforts to provide the full technical support, resources, and training to the instructors of the higher educational institutions to improve the adoption of LMS. The findings and suggestions of this study would allow MoE and higher educational institutions management to re-evaluate their existing practices and support systems for LMS, and if there are issues that need to be addressed.

2. **Making Technological Teaching and Resource Centers**: The majority of the instructors agreed that the management is focusing on the technical support of the students. There is no dedicated technical support available for the teaching staff. In case of any problems, the same IT staff come and try to resolve the technical problems. Due to this trial and error troubleshooting, the teaching and learning process suffers. Hence, besides regular technical support centers, it is recommended that well-resourced technical support centers for teaching staff be developed. The recommended technical support center should have a high-bandwidth Internet connection through which multi-disciplinary pedagogical materials can be accessed. It should provide training to troubleshoot simple technology-related issues. It should also have online consulting to assist instructors with a variety of instructional technology and technical support. Walk-in help centers to promptly troubleshoot technical problems are also recommended. In this regard, it is recommended to make a liaison between coordinator positions to build bridges between technical staff, academic departments, instructors, and students. It is also recommended that instructors could be motivated, compensated, and supported with incentives such as hand-held projectors and laptops for their seamless content delivery.

3. **Addressing instructors' uncertainty situations (such as fears about safety, security, and accessibility) of stored data**: The concerns regarding safety, security, accessibility of teaching material, and stored assessments continue to be the key fears of the instructors in LMS adoption. The management of HEIs should build the trust of instructors regarding technology by investing in technology improvement, data privacy, and security. The LMS providers should incorporate modern technological advances to secure the data in the LMS to make instructors more comfortable and to have more confidence in using technology. LMS providers, MoE, and the management of the HEIs should also evaluate the legal issues related to data privacy, and its security to build

trust among the instructors to encourage them to use LMS in their teaching.

4. **Consideration of Usefulness and Comfortability Component of LMS**: The institutions should make greater use of LMSs, which are valuable platforms for sharing materials, content delivery, collaboration, communication with peers and students, and participating in forums. The management of HEIs should make sure that the usefulness and benefits of the LMS are given due attention. In light of such a finding, it is recommended that the usefulness of LMSs should be an important factor for LMS designers to consider when developing LMS-related applications. Furthermore, effort expectancy (ease of use) being a significant predictor of LMS usage indicates that instructors having positive perceptions on the ease of use and comfortability with the LMS would have greater intention to adopt the LMS. Hence, LMS designers are recommended to use the findings of this study to reassess and redesign the interface structure of the LMS. It is recommended that unnecessary complexity should be removed from the LMS interface to make it simple. To promote their product, the LMS providers may need to ensure that their systems are easy-to-use, user-friendly, entertaining, and compatible with all teaching and learning environments.

5. **Improving Literacy and Awareness of Benefits of the LMS**: The findings of this study suggest the need to raise awareness of the benefits of LMS among instructors. It can be highlighted through the mass media or via interpersonal levels (Rogers, 1995). Management might consider investing in the advertisement to improve literacy and the awareness of the LMS among instructors at higher educational institutions and encourage those instructors who do use the LMS to convey a positive message to those who do not use it.

6. **Choosing Right LMS and Right LMS Provider**: Blackboard was introduced by Blackboard Corporation in 1997 and is being used for teaching and learning in worldwide universities (Sharma et al., 2011). As technology continues to advance, Blackboard experiences growing competition from competitors, such as Moodle, Desire2learn, and other rivals. Due to the availability of a diverse LMSs, higher educational institutions now have a wide variety of options to choose from to manage their learning curriculum. Each LMS provides specific functionalities and content management approaches, so choosing the appropriate LMS system becomes an important concern for the management of higher educational institutions. Another concern is that one LMS does not provide enough finalized, comprehensive functions to

satisfy all the demands of the institutions. Therefore, it is not possible to recommend a specific LMS for all institutions. The key features of the LMSs are the consideration for the management to choose a flexible LMS system that fits the teaching and learning environments of their institution. The management should be aware of the importance of not only the right choice of the LMS but also selecting the trusted providers of the LMS. The right LMS should fit into the teaching properly and the right LMS providers should provide timely and appropriate technical support when needed.

7. **Considering Cultural Aspects**: In the Middle East culture, religion plays an important role in shaping social norms, traditions, obligations, and practices of society (Al-saggaf, 2004). To bring increased acceptance of the new technology, it is suggested that technology should be designed to match with the cultural values of the society. Since the culture strongly moderates the adoption behavior of the individuals within the society, the practitioners and policymakers should consider that the implementation may be tied to acceptance within these specific cultural parameters.

8. **Need for Development of Advance Features of LMS**

- Mobile applications are currently available in most LMSs. LMSs should also include a text-based interface so that off-campus students having a low level of Internet connectivity can also access the content. In some areas, the loading speed of LMS becomes a barrier to access, due to low bandwidth. The existence of a text-based interface will increase the accessibility of the course sites. Therefore, it is a recommendation for LMS designers that options for lighter applications should be made available so that off-campus students may reap the benefits of LMS.

- Although standard search features are available in most LMSs, advance searchable features are not available in some cases. It is recommended that search systems within a course site should be improved by integrating enhanced search features in LMSs. Such improvements would be beneficial for instructors and students, as well.

- Students who are not aware of technical issues may be vulnerable to security risks such as phishing, theft, and computer viruses. There should be a text or sound warning to the students leaving without logging out. It is important to note that a text-based warning about critical risks might be overlooked by students. Therefore, critical situations could be stressed by the use of sounds or images, such as a pop-up window with a sound to warn students that they are about to leave the LMS site. This would produce an error-tolerant system.

9.5.2 Micro-Level: Recommendations for Instructors and Researchers

(a) **Recommendations for Instructors**

- **Participating in Professional Development Programs**: It was indicated by instructors that there is a need for training on the new Blackboard system because they were unaware of many features of the new Blackboard version. Thus, such instructors will perceive technology to be hard to use, resulting in the production of avoidance behaviors concerning (LMS) technology used for teaching and learning. Professional development programs of an LMS tailored to new and more experienced instructors are recommended at the start of every academic year. The participation of instructors in professional development programs could enhance instructors' skills, knowledge, alternative pedagogical strategies, and emerging educational tools such as LMSs (Doutrich, Hoeskel, Wykoff, & Thiele, 2005).

- **Communicating and Interacting Using an LMS**: It is clear from the results that the majority of the instructors consider that an LMS is the most suitable interaction tool for communication with students. The findings indicate that improved communications with chat, emails, and mobile apps of LMS appeared to impact positively on the progress of the students. Instructors' accessibility and feedback is an essential part of LMS usage for the students. It is recommended that instructors should make contact with students and stay involved. To make the better use of LMS, course instructors should be engaged in the discussion forums, ask questions, clarify their doubts, and focus the discussion on important topics.

- **Using Higher Level Features of LMS**: The literature on LMS shows that various higher level features (such as lesson activity module, OU Wiki activities, Blackboard collaboration, Blackboard Instructors, and Blackboard Apps for students) are available in the recent LMSs which are currently underutilized. The instructors are recommended to use the higher level features of LMSs in their teaching. The instructors should also consider using assistive technologies in their teaching. Assistive technologies, for instance, voice recognition would assist instructors and students in the teaching and learning process. It would be beneficial not only for instructors but also for the students.

- **Making Learning Communities**: In teaching and learning environments instructors are expected to interact with each other, discuss issues, share the experience of their students, and exchange scholarly work. It is recommended that 'learning communities' for the instructors should be formed within the institutions to nurture collaborative learning and teaching practices. Learning communities would help in improving trust, interaction, and sharing of knowledge of the instructors (Nett, 2008). It can be any mode of interaction such as face-to-face, social media, or a combination of both. The instructors can meet regularly to discuss and update their skills and knowledge on LMS technology, share ideas on current research, and can also restructure the courses they are teaching.

- **Recommending Peers and Students**: The results of this research lead to the inference that ease of use of an LMS is an important factor that influences the behavioral intentions of instructors to adopt an LMS. It is recommended that instructors with positive perceptions of LMS should recommend and encourage other instructors to use LMS in their teaching.

(b) **Recommendations for researchers**

- **Removing the Construct 'Habit' from original UTAUT2 model**: The original UTAUT2 model used a 'habit' construct in the context of mobile phones. In this study, the construct 'habit' was removed due to its poor reliability and validity. It was found that individuals are more dependent on mobile phones than on LMS, as LMS is only a teaching and learning tool. Since this construct is not relevant to LMS, the researchers in the area of LMS adoption are recommended to rephrase the questions about this construct by asking specifically if the 'LMS has become their habit in teaching' or by removing this construct.

- **Considering Demographic Variables**: Although professional assistance, institutional support, and cultural factors influence the adoption of LMS, a clear picture of demographic variables such as differences in personality factors, academic discipline, curriculum, association, and institutional policies, is also required. Researchers are recommended to consider all possible demographic influential variables.

9.6 Summary

This study developed and employed an extended UTAUT2 model to assess the factors that influenced the adoption of LMS at higher educational institutions of Saudi Arabia. The results showed that the constructs PE, EE, FC, SI, and HM (except H) were significant and were strong predictors in the adoption of LMS. This study attempted to address the limitations of the original UTAUT2 model and has developed an amalgamated model by incorporating new (moderating) variables such as cultural dimensions and technology awareness. It is expected that the proposed model will serve as a guide for the management of higher educational institutions considering the adoption of LMS at their institutions. The literature is lacking in the implementation of the original UTAUT2 model in LMS adoption by instructors in the cultural context of higher educational institutions. The new aggregated technology adoption model may be considered as a transferable model that can be applied to test instructors' behavioral intentions in other regions. Thus, the proposed model provides a new methodology, fills gaps in the literature, and thus reflects an effort to expand the UTAUT2 model so that it can be applied to other educational institutions in different regions. Due to cultural similarities between Saudi Arabia and other Gulf countries such as Jordon, Kuwait, and Qatar, the results can also be generalized to other GCC countries. Furthermore, the enhanced questionnaire developed for this study offers a unique instrument for researchers in the adoption of LMS in non-Western countries. Hence, this is the first study of its kind which applied the extended UTAUT2 model in exploring instructors' intention to use LMS in higher educational institutions of Saudi Arabia and is expected to be applicable in higher educational institutions of other GCC countries.

References

Al-saggaf, Y. (2004). The effect of online community on offline community in Saudi Arabia. *Ejisdc*, 1–16.

Al-Gahtani, S. S., Hubona, G. S., & Wang, J. (2007). Information technology (IT) in Saudi Arabia: Culture and the acceptance and use of IT. *Information & Management, 44*(8), 681–691. https://doi.org/10.1016/j.im.2007.09.002.

Davis, F. D. (1989). Perceived usefulness, perceived ease of use, and user acceptance of information technology. *MIS Quarterly*, 319–340.

Doutrich, D., Hoeskel, R., Wykoff, L., & Thiele, J. (2005). No Title. *Teaching Teachers to Teach with Technology, 36*(1), 25–31.

Fishbein, M., & Ajzen, I. (1975). *Belief, attitude, intention, and behavior: An introduction to theory and research.* Don Mills, Ontario: Addison-Wesley Publishing Company.

Hofstede, G. (1980). Culture and organizations. *International Studies of Management & Organization, 10*(4), 15–41.

Nett, B. (2008). A community of practice among tutors enabling student participation in a seminar preparation. *Computer-Supported Collaborative Learning, 3*, 53–67.

Rogers, E. M. (1995). *Diffusion of innovations* (12th ed.).

Sharma, S. A. T., Paul, A., Gillies, D., Conway, C., Nesbitt, S., Ripstein, I. R. A., Simon, I., & Mcconnell, K. (2011). Learning/Curriculum Management Systems (LCMS): Emergence of a new wave in medical education. *Learning, 11*(13).

Tosunta, B., Karada, E., & Orhan, S. (2015). The factors affecting acceptance and use of interactive whiteboard within the scope of FATIH project: A structural equation model based on the Unified Theory of acceptance and use of technology. *Computers & Education, 81*, 169–178. https://doi.org/10.1016/j.compedu.2014.10.009.

Venkatesh, V., Morris, M. G., Davis, G. B., & Davis, F. D. (2003). User acceptance of information technology: Toward a unified view. *27*(3), 425–478.

Venkatesh, V., Thong, J., & Xu, X. (2012). Consumer acceptance and user of information technology: Extending the unified theory of acceptance and use of technology. *MIS Quarterly, 36*(1), 157–178.